ARTHUR H. BELL

Mastering the Meeting Maze

ADDISON-WESLEY PUBLISHING COMPANY, INC.

Reading, Massachusetts Menlo Park, California New York
Don Mills, Ontario Wokingham, England Amsterdam Bonn
Sydney Singapore Tokyo Madrid San Juan

Many of the designations used by manufacturers and sellers to distinguish their products are claimed as trademarks. Where those designations appear in this book and Addison-Wesley was aware of the trademark claim, the designations have been printed in capital letters (e.g., Apple Computers).

Library of Congress Cataloging-in-Publication Data

Bell, Arthur H. (Arthur Henry), 1946–
Mastering the meeting maze / Arthur H. Bell.
p. cm.
Includes bibliographical references.
ISBN 0-201-55064-4
ISBN 0-201-52362-0 (pbk.)
1. Meetings. 2. Business communication. I. Title
HF5734.5.B45 1990 89-18302
658.4'56—dc20 CIP

Cover design by Mike Fender
Text design by Joyce C. Weston
Set in 11-point Bookman by Pine Tree Composition, Lewiston, ME

ABCDEFGHIJ-MW-9543210
First printing, February 1990

Contents

Acknowledgments

TO the hundreds of business men and women who share their meeting experiences and insights in this book, I am deeply grateful. For generous permission to reprint several cartoons, I thank editor Kerry Rottenberger of *Sales and Marketing Management* and the magazine's parent company, Bill Communications. Finally, I want to express appreciation to a superb editorial team at Addison-Wesley: Kate Habib, Kathy Manley, Sandy Hutchinson, Perry McIntosh, and their colleagues.

For my parents with love, Art and Dorothy Bell.

Foreword

WHEN Art Bell phoned to invite me to write a foreword for *Mastering the Meeting Maze,* I was . . . in a meeting. American executives can too easily get caught up in a web of meetings, with no free time to think, to listen, and to act.

This book essentially asks two basic questions: *Which meetings are really necessary,* and *how do you get the most out of them?* Any executive who can answer those two questions can save countless hours of wasted meeting time and thousands of dollars of misdirected resources.

And it's not that hard to have fewer, better meetings. As these chapters show, many of our wasted meetings are simply bad habits. Managers have often gotten used to sitting around a table a certain number of hours each day. In time, they come to think that's what it means to "do business."

It's always amazed me that business schools don't teach what every executive finds himself or herself doing each day: planning meetings, running meetings, participating in meetings, and following up after meetings. *Mastering the Meeting Maze* fills that gap. This book is a lively, street-wise course in how to make meetings matter.

ARTHUR POTTASH
Chief Executive Officer
ARTEX Knitting Mills, Inc.

Introduction

WE'VE all been there. Eight, ten or more of us barely stifle yawns as the regular talkers hold forth. An hour into yet another meeting, with no end in sight. And no agenda.

Any questions on that point? There are none, but Herb thinks we ought to talk about it anyway. Actually, Herb thinks we ought to listen to *him* talk about it. *Thanks for your contribution, Herb. Anyone else have anything?*

Most of us do, of course: a pile of *work* sitting on our desks—work we could have completed *at* work, had it not been for meetings, meetings, meetings.

American business meetings come in red, white, and blue.

Red meetings occur when Pat, the boss, perceives an emergency of some kind. Red alert! Drop everything—you *will* rearrange your schedule to be there. Ten of us meet. Pat stands up front, red-eyed. But the emergency turns out to be only a brush fire, and is quickly doused. One person in the group has an answer for the boss. The two of them could have talked privately.

But no. It suited the boss, red-faced now, to take ten people away from their work for an hour of . . . well, mismanagement.

The truth is that the boss *likes* red meetings. Pat gets a chance to bluster, to actually wrestle out loud with a problem. This is the boss in action. The fearless leader creates a moment of drama. Or soap opera—trite, easily resolved plots and problems with the same predictable cast of characters.

White meetings are neutral and routine. On Tuesday morning, for example, we regularly attend the 10 o'clock meeting. Why hold the Tuesday meeting? Because it's *Tuesday.* Why do the ten of us attend? Because we *all* attend the Tuesday meeting. And what do we do with the best work hours of Tuesday morning? Well . . . *meet.*

White meetings have one goal: to *discover* the goal of white meetings—to see if anyone "has anything." Good sports all, we dutifully take our turns at bat—little problems we've already solved (making us look good in the telling) or big problems that we can't possibly solve in just one meeting. It's a good time to catch up on snide remarks about the current company goat.

The agenda for white meetings is always short and vague. Items listed for discussion begin with "Tentative planning for . . ." and "Preliminary ideas regarding . . . ." The discussion is also short and vague. *What do you have on that, Frank?* Not much, of course, since Frank didn't know he would be called on and therefore left his files in his office. *Well, does anyone have something?* Uh . . . no.

White meetings produce *white-out,* that peculiar blindness to the obvious. Eight, ten, or more people regularly enact a ritual that gives the illusion of work being done, business being transacted. Behind the curtains lies the truth: play-actors doing little good for themselves, the company, or the clients.

Sadly, veteran employees often become snowblind to this travesty. After years of wandering through white meetings, they lose sight of wasted time and misspent money. They can't imagine a work-week without white meetings.

Blue meetings occur when the boss is blue. It's a slow Thursday and the boss's lunch appointment just cancelled. Hell, why not get the group together? Find out *what's happening.* Share some *perspectives.* Bring each other *up to date.*

Blue meetings come out of the blue. We arrive without a

clue as to why we were called away from our work. Oh, it's a *blue* meeting, we recognize once we get there: the boss is stalling, chatting with one individual, then another . . . leaving the room now to "check on something." Like strangers at a cocktail party, we settle to the task of making chit-chat. The boss returns, the blue mood showing.

Bosses, after all, are social creatures. They got out of labs and took off green eye shades because they wanted to manage people—to be with people, to impose themselves on people, to be thought about and somewhat feared like an in-law. When the boss's business life gets dreary, when the phone doesn't ring—*where are the children when you need them?*

So we gather round. The business of a blue meeting is essentially ego massage. No matter what the topic, as "perspectives" and "feedback" are given around the table the deep communication is the same: "We think you're swell, boss. Don't worry. Be happy."

Enough!

If the chiefs can't see the huge percentage of waste in red, white, and blue meetings, the Indians should tell them. Here are the raw facts about meetings as America prepares for the global economic struggles of the 1990s and beyond:

We're meeting more, not less. The electronic communication revolution has freed us from many time-consuming handtasks (calculation, old-fashioned typing, carbons, and slow photocopying) so that we have more time. For what? Well . . . meetings.

We're still—bosses and workers—essentially untrained in how to conduct or participate in a meeting. Less than 10 percent of students graduating from American business schools spend any class time at all on meeting skills. An even lower percentage of in-house corporate training programs conduct sessions on effective meetings.

We're spending more than ever on meetings. A two-hour meeting occupying the time of six mid-level managers runs to $1200 in salaried time alone. Across industries, as much as fifty percent of time spent in meetings is wasted—that is, absolutely without justification by any work-related rationale.

We're fed up with meetings. All surveys of employee attitudes since 1985 show worker malaise and increased hostility toward hours spent in meetings versus hours spent at real work tasks. Meetings are frequently listed by workers as their least favorite and least useful work activity.

What to Do

Informed by the best meeting techniques from recent management literature and by fifteen years of consulting in meeting strategies for major companies, I suggest in this book a practical approach to fewer, more productive meetings. *Fortune, Forbes, Business Week, Harvard Business Review, Inc.*, and other business oracles have told us for years that American business is undergoing dramatic changes as we near the end of the century. Together, we hope we're entering upon the spring of American prosperity, not upon winter—the budding of ideas, a time of growth, an era of new birth.

The greening of American business should color meetings as well.

1

To Meet or Not to Meet

HOW many meetings are held each year in American business? Ask any manager and you'll get the same answer: *too many!* (For trivia buffs, the actual number exceeds the country's population—350 million or more.) Today, American managers are spending about half their time sitting in meetings.

Are many or even most of those hours wasted? Joan Hamann, a director at the University of Wisconsin's Management Institute, thinks so: "Talk to any group involved in a time-management seminar. They will overwhelmingly condemn meetings as the single biggest time-waster in their work worlds. There is an almost universal dislike and frustration associated with meetings, especially among productive, results-oriented managers."[1]

Peter Drucker, as usual, states the case even more strongly: "You can work or you can meet. You can't do both at the same time." For Drucker, a manager's spending more than 25 percent of his or her time in meetings is a sign of "malorganization."

But I suspect you're interested in your own flat tire, not the state of the rubber industry. This book isn't a Kinsey report on the sad state of meetings in America. Rather, it's a collection of ideas about what may be wrong with your meetings, and what you can do about it.

---

[1]All references are listed alphabetically on pp. 175–179 ff.

THE TICKING CLOCK

Andrew Leigh of the Institute of Personnel Management in London tells the story of a CEO who "recently opened a meeting of his managers by placing an oversized digital clock on the conference table where everyone could see it. He paused to let the effect sink in and to allow the participants to notice that the machine was not showing the correct time.

"With a faint smile, the chairman announced: 'This clock shows the cost of this meeting in terms of dollars per minute. As you can see, we've so far spent just under $500.' "

The remainder of the meeting, Leigh reports, was conducted "at an unusually vigorous pace."

That same clock ticks for each of your meetings. At Intel Corp., for example, a mid-level manager's time is valued at about $100 per hour. Intel president Andrew S. Grove does the calculation: "A meeting attended by ten managers for two hours thus costs the company $2,000."

What disturbs Grove is that "most expenditures of $2,000 —such as buying a copying machine or making a transatlantic trip—have to be approved in advance by senior people, yet a manager can call a meeting and commit $2,000 worth of managerial resources on a whim. If that meeting is unnecessary or so poorly run that it achieves nothing, that's $2,000 wasted."

Last May, six senior managers at a major insurance company engaged in three prolonged meetings to determine whether to buy a 12-seat commuter van or a somewhat more expensive, and more comfortable, 16-seat van. They decided, conservatively, in favor of the less expensive van and recommended its purchase to a company vice president. To their chagrin, she shot back an answer: "The money you spent in meetings would have allowed us to buy the larger van. Think about that as you squeeze into your 12-seater on the way to work!"

After 30 years with McDonnell Douglas, senior executive Jerry Patterson reflects that "there were probably at least ten meetings with ten or more people going on all the time at our location. That's 800 man-hours a day, or 208,000 man-hours a year. If Uncle Sam pays $100 an hour for engineering, that's over $20 million per year at one plant for meetings. The question is whether they move the programs ahead enough to pay for themselves, and my answer, generally, is no."

The real cost of bad meetings can't be computed simply by measuring managerial time. There's also an attitudinal or emotional clock ticking—the kind of time bomb that makes bright minds burn out and sends good employees to the competition.

What can it mean for a company, after all, when 70 percent of its managers find meetings "a waste of time" (the result of a Communispond survey of 471 managers)? Creative, goal-oriented people don't willingly endure work environments where, as *Datamation* says of meetings, "the boastful can find listeners, the aggressive can find victims, and the indolent and the slothful can find repose."

As exit interviews across industries show, good people quit in droves when they feel they're wasting their time—when meeting after meeting becomes, in John A. Lincoln's phrase, "a cul de sac where ideas are lured and quietly strangled."

THE BLAME FOR BAD MEETINGS

To identify a perpetrator for the crime of bad meetings, we have a rogue's gallery of likely possibilities.

Suspect 1: Lack of Preparation

Almost 90 percent of male and female executives queried in a poll reported in *Working Woman* in 1987 believe that "business meetings fail because of lack of preparation." But what is "lack of preparation" for meetings? I asked that question of 20 managers and received 20 different answers, including these:

- ▲ no sense of direction
- ▲ a total lack of control

- ▲ an inability to state the purpose of the meeting
- ▲ a failure to get down to work in the meeting
- ▲ no clear idea of what we were supposed to do
- ▲ no agenda
- ▲ no explanation of why the meeting participants were chosen to be there

Such broad and varied responses confirmed my suspicion about the 90 percent answer, "lack of preparation." That phrase was more a *gesture of frustration* by executives than a precise description of what, in fact, was going wrong in meetings. Like the word "communication," the phrase "lack of preparation" meant so much as to mean almost nothing.

In one manager's office I was shown a terse memo from a vice president of operations: "The Executive Committee has noted a tendency toward inefficiency in meetings. I urge each of you to be more thorough in your preparation for these occasions."

There's that word again—*preparation.* But has the vice president changed anything by this solemn advice? Will any manager in the company know what to do, or not to do, based on this message from on high?

Probably not. The great majority of managers—78 percent, in a survey reported in *Personnel Journal*—indicate that they have received no formal training in how to conduct or participate in a business meeting. And it doesn't just come naturally, says Patti Munzer, Celanese Fibers Inc.'s director of marketing communications: "I wasn't born knowing how to lead meetings."

When no one is driving the bus and none of the passengers are in their seats, the result is a bumpy ride indeed. Stanley Bing remembers hearing of a case in point:

> "A marketeer I know recently found himself trapped in an all-day quarterly planning meeting. Crammed like clowns into a Volkswagen bug, the executive team of thirty were to float their ideas past the chairman. 'Fear subsumed the room,' he reports. 'Everybody was there for two reasons: to launch their balloons and to shoot down everyone else's. I stood up

and gave what I thought was a very good idea, and there was this little pause, then the chairman said, 'Okay. On to the next!' I was devastated. That sort of thing went on all day. I haven't had so much fun since I was stabbed in the head."

Only in the broadest sense can we blame "lack of preparation" for such a debacle. What this chairman would have to learn to truly *prepare* for a productive meeting goes far beyond a better agenda or improved ordering of topics. This curmudgeon needs the full body wrap, the Golden Door make-over, the head-to-toe attitudinal rehab to ever function productively in meetings. Coming from this fire-storm environment, meeting participants will need help of their own—almost therapy. They will need to learn to value one another's ideas, to express themselves without fear, and to listen.

Suspect 2: Too Many Meetings Spoil All Meetings

"Most managers take it as a point of honor to dislike meetings, at least meetings called by somebody else," writes Walter Kiechell III in *Fortune*. "'If we didn't have so damn many,'" the standard complaint runs, "'we might actually get a little work done.'"

Count the meetings, large and small, you've attended in the last 30 days. Many managers discover that they meet more often than they eat. John Cleese, of *A Fish Called Wanda* fame, has produced and starred in two minor classics aimed at business people: "Meetings, Bloody Meetings" and "More Bloody Meetings" (available through Video Arts). In one scene from the first of these, Cleese seats himself at the head of a conference table to chair the Monday meeting. Why meet on Monday? Because, Cleese explains, it's the usual Monday meeting. We always meet on Monday. And what if the group has nothing to discuss or decide? Well, Cleese retorts in perfect deadpan, we won't know that if we don't meet.

Entrepreneurs scoff at the endless hours corporations devote to ritualistic meetings. "I don't go to unnecessary meetings,'' says Gordon Schmidt, owner of Daybreak Holsteins. ''It's one advantage of being self-employed.''

Meetings have become the worn but accepted furniture of business life—the stuffed chairs where thinking takes place, ideas are mulled over, and social life is pursued. Meetings are so omnipresent that the business day would seem empty somehow without two or three of them. Things are a bit slow this afternoon. Shall we meet?

Not if you have a job to get done. Susan Guggenheim Lodge, publisher of *M* magazine, has cleared the decks of unnecessary meetings in her organization. She keeps in-house meetings to an absolute minimum because ''our sales staff's time is more profitably spent on the telephone.'' Lodge recognizes that meetings are too often a comfortable alternative to *doing,* in the same way that government panels are a common substitute for action that resolves social problems.

Perhaps the most maddening of wasted meetings are those devoted simply to information transfer. Ward Hemer, engineering manager at Sperry, recalls ''a number of meetings that were just announcements. The boss could have had a secretary send us a note.'' At Eastman Kodak, says technical engineer James Lindbloom, ''75 percent of all the meetings I've attended could have been covered in a three- or four-paragraph handout.''

Sometimes such information-''sharing'' meetings are pure sham for both presenters and audience. ''I was called to a meeting to present information on a policy change,'' recounts Dan Mackensen, senior research scientist with Hybritech. ''The change had already been cut in stone and response from the group was not being solicited. The information covered in the meeting could just as well have been communicated through a single memo.''

''If you can get the job done without holding a meeting,'' says Nido Qubein in *Communicate Like a Pro,* ''then calling one is a waste of time.'' That notion may seem revolutionary to managers who have always thought of themselves as meeting arrangers rather than work producers. But have no fear. No group of employees has ever stormed the corporate

Bastille crying for "more meetings, more meetings!" In fact, says Geraldine Henze, communications director at Columbia School of Business, "you rarely lose authority by calling too few meetings. You gain the respect of your co-workers and subordinates by showing you value their time."

Employees, after all, can read. At Bull HN Information Systems, education consultant John Draftz resents meetings spent listening to committee reports. "Why not publish the various reports, give them to me, and forget the meeting?" he asks. "I can read faster than they can speak and a phone call can clarify any questions." Meetings that are no more than listening sessions make employees feel "insignificant," says Draftz.

As a rule of thumb, *half* as many meetings would be *twice* as good. Each would have more real work to do, more real issues on the agenda, more real energy in discussion and debate. As in art, so in meetings, the Bauhaus principle: "Less is more."

Suspect 3: When in Doubt, We Hold a Meeting

Take this test, suggests a recent issue of *Forbes*. "You have to come up with a marketing strategy for the hottest new product in your division. You first move is to

a. decide what's to be done and write your boss a memo
b. talk it over privately with two associates and then write the memo
c. call a meeting and get everyone involved."

When I gave this quick quiz to 50 M.B.A. students with several years of corporate experience, I received a schizoid response. While most respondents felt that *c* was certainly the least advisable choice, they admitted that it was by far the most likely choice in the organizations from which they had come. Meetings, after all, are a superb smokescreen for both inactivity and individual responsibility. Why isn't Project X on its way to production? "We're meeting tomorrow about that." When will the report be written? "We're discussing that at Tuesday's meeting."

Such procrastination and obfuscation infect many organizations at all levels. Even top executives resort to hiding behind their company's meetings, says Beverly Hyman, a well-known meeting consultant. "If a decision doesn't work out, the executive in charge can always say, 'We all agreed at the meeting, didn't we?'"

Few business people doubt the *U.S. News and World Report* estimate that "25 percent to 50 percent of all meetings [are] unnecessary or unproductive." But the implications for a manager spending half of his or her day in meetings are devastating: two or three hours of each day lost to meaningless, often redundant activity. Isadore Barmash has gone so far as to suggest, albeit facetiously, that American productivity could be measured in relation to the "trillions of man- or woman-hours annually that can be translated into lost productivity" due to meetings. The American Meetings Rate, he proposes, could be considered a new economic indicator. The fewer the meetings, the higher the GNP.

SO MEETINGS ARE IN TROUBLE—WHAT'S THE GOOD NEWS?

To say that we meet too often and too carelessly is not to say that we should not meet at all. At their best, meetings are useful managerial tools for

Getting out information in a simultaneous way to a select audience, with opportunity for on-the-spot questions and comments. (A memo containing the same information might be lengthy, subject to misinterpretation, and grist for the rumor-mill—half-truths that may well reach others in the grapevine before they read the memo itself.)

Demonstrating and modeling the cooperative, interdependent work relations you want to foster in the company. George in Accounting may work better with Linda in Purchasing because the two participated in a meeting where their ideas were encouraged and accepted.

Gathering and sharing input necessary for decision-making. (While input can also be collected by memos and reports, much business information is relational in nature. Bob can't speak definitively about his data until he knows what Gloria plans to do about hiring, which in turn depends upon budget figures from Susan, and so forth. These connections can be made quickly and naturally in a meeting, but may consume reams of paper if attempted by written channels.)

These rather tame defenses of meetings are the velvet glove covering other iron realities of meetings. Although the agenda never says so, meetings are the arena in which, as Joe Kelly writes in *Personnel*, you are "coolly appraised by reigning members of the management group who are deciding whether to let you in." At the same time, the meeting group celebrates and reinforces itself. "The meeting is essentially a tribal assembly in which the attending managers are reassured that they are 'on the team'," suggests Kelly.

Meetings also provide, usually in an unstructured and unannounced way, a forum for public praise and public whipping. An appreciative comment from the chair after Ruth's report may motivate her more than a raise. On the other hand, the chair's dispeptic or inattentive look during Al's presentation rouses the blood-thirst of the group: 'Let's crucify the S.O.B.'"

Meetings are a stage for other forms of *drama* and *trauma*. Conflict among peers, for example, can be exercised and exorcised within social limits. Sparring takes place in exquisitely subtle ways: an interruption, a terse response, a pointed joke, a tone of voice, a facial expression. The goal of such encounters, says Kelly, is "not necessarily to be right; rather, one must sound convincing and denigrate one's opponent as coolly and elegantly as possible." In the Kennedy-Nixon presidential debates, for example, Kennedy adopted what Kelly calls "a certain patrician aloofness that made Nixon look shifty by comparison."

In short, meetings remain a useful tool for both **overt management** (team building, information flow, decision-making, and motivation) and **covert management** (evaluation, drama, and trauma).

THE MANAGEMENT OF MEETINGS

"I can call the spirits from the vasty deep," wrote Shakespeare, "but will they *come*?" Your ability to harness the strengths of meetings without stepping into trouble depends on how you manage the beast.

Begin by asking yourself *what you want*—in effect, the objectives of the meeting. Carl Huffman, president of Huffman Enterprises, Inc., has a commonsense approach: "Objectives are what the presenter wants the audience to think, feel, and do." Sometimes your objective may be quite specific: "I want the group to support the X2 Project by a consensus vote." At other times, the objective may be more general: "I want each member of the group to have a chance to ask questions about the Financial Disclosure provisions in the new contracts." Whatever the objectives, you must grasp them clearly before you can organize a meeting to accomplish them.

Making Your Purpose Clear

If your path becomes clearer once you know your objectives, why not also enlighten group members by telling *them* the purpose of the meeting? This obvious first step in meeting management is often countered by the rejoinder, "I'll fill them in at the meeting." Meanwhile, your co-workers or subordinates may speculate and worry about why the meeting was called, may spend time preparing for topics that won't even be discussed at the meeting, and may find reasons not to attend.

The purpose of the meeting is best stated in the initial meeting announcement: "We will meet for one hour to review contract specifications for the Richfield project and to make a final selection of a design architect." There, that didn't hurt. Attendees know *why* they must be there, *what* they will be expected to do, and *how long* it will take.

Managing meetings means managing time. Andrew Crighton, executive vice-president of Communique Ltd. of Toronto, keeps tight rein on time in meetings he conducts.

"If a meeting doesn't start on time, it isn't really being treated as a formal business meeting. It's just a coffee break or gathering." Show up late at one of Crighton's meetings and you'll find the train has left. He doesn't hesitate to start without the stragglers as a way to "try to make them feel guilty."

Even more electric is the shock applied to latecomers by an Argentine industrialist. Shortly after his meetings begin, he has an aide remove the chairs of anyone who has not arrived. Often they do finally arrive—and stand.

Group members will appreciate your firm management of time—*their* time. "Everyone has to respect everyone else's time," says Goldie Brass, national director of meeting planning for Shoppers Drug Mart. "Five high-powered people should not be kept waiting for the sixth to arrive. Yet it happens all the time."

Policies for Meetings

You can extend your influence over meetings in your organization by establishing meeting policies. The employee handbook or policy manual probably contains specific guidelines for rather trivial matters: coffee breaks, supplies, and parking. Why not policies on meetings? The goal of such policies would be to set forth what the company expects from meetings, how they should be held, and for what purpose. A Texas computer manufacturer, for example, specifies by policy that an agenda must be distributed at least two days prior to a scheduled meeting; that the agenda must be filed in advance along with a brief justification for the meeting with the executive vice president; and, that the outcome of the meeting be reported to all stakeholders in clear, concise minutes. "We meet less," says a company engineer, "because we have to think about why we are meeting in advance. Besides, it just isn't easy to have a meeting around here."

Specific suggestions for managing meetings are the business of the following chapters. But, for now, consider this humorously prophetic view of meeting management that appeared in the *Newsletter for Xerox Managers*:

"And as the conclaves multiplied and heaped upon one another, they were delayed from starting and delayed in ending and were postponed to be called again when those whose presence was required could be made available. For the message was clear, but the ways were hidden.

Then from all sides came voices crying:

'Though I labor from my coming in to my going out, I cannot attend the meetings for which I am summoned.'

'It concerned me not, yet I was called unto a meeting.'

'My need for decision is great, yet am I denied, for all are in meetings.'

'Can the meeting not start by the mark on the hourglass?'

'To the meeting for which I made ready, no person came.'

'Is thy servant a fool, that thou summonest again to a meeting to schedule meetings?'

And then at last the chief, hearing of these things, decreed that the ax be laid to the root of those meetings which brought forth bad fruits. And he sent forth a scroll, saying

I. Thou shalt not meet if the matter can be resolved by other means.
II. Thou shalt make the purpose of each meeting known to those thou summonest.
III. Thou shalt summon only those whose presence is needed.
IV. Thou shalt start at the time announced.
V. Thou shalt stop when it is meet and right so to do.
VI. Thou shalt not run beyond.
VII. Thou shouldst combine into one those which need not be separate.
VIII. Prepare thy thought, that the minutes be not wasted.
IX. Schedule not in haste, for the day is short in which to do that which thou hast to do.
X. Fear not to cancel if need disappears.

And in time, the people learned and obeyed these writings and followed them, putting to and taking away as suited their needs. For as they forsook their old ways, new hours were given unto them, and they were free to do their things, and they saw that it was good."

ROLODEX RECAP

▲ unnecessary meetings are expensive in lost dollars and sabotaged morale

▲ bad meetings can be blamed on poor preparation, redundancy, and lack of focus

▲ good meetings, in an overt way, serve the valuable functions of getting out information, modeling interdependent work relations, and gathering/sharing input

ROLODEX RECAP

▲ good meetings, in a covert way, provide an opportunity for personnel evaluation, hierarchy reinforcement, and conflict resolution

2

What to Expect in Meetings

LIKE tourists at Universal Studios, many meeting leaders and attendees aren't sure what set they're visiting and what's supposed to happen there. David Corvo, executive producer of *CBS This Morning*, is "stunned by the number of people who don't know what kind of meeting they're in. One of the best examples is when the boss convenes a meeting to convey a decision and a staff member persists in debating the decision. The time for that is past!"

Different kinds of meetings? That idea comes as news to managers who believe they have experienced a sausage-string of meetings in their careers—meetings all alike in taste, color, and lack of nutritional value. But, in the immortal words of Jimmy Dean, "not all sausage is alike."

THE MISSION MEETING

At least half of all business meetings are called with a mission in mind—to make a decision, to solve a problem, to formulate policy. These meetings are often convened quickly, with little time for participants to gather information and prepare remarks. Attendees are usually stakeholders in the issue at hand: those who must approve funding, conduct marketing, or supervise manufacturing.

"I can tell incredible stories about people who didn't know they were at a mission meeting," says an aerospace executive. "My company was in hot competition for a large Navy contract and the boss called an emergency meeting to iron out some wrinkles. For an update on a couple of technical points, he called on one guy from Electrical Inspection, who stood up and proceeded to tell us the history of his division, including his urgent need for a new secretary. The guy might just as well have climbed into his own coffin."

The rules for mission meetings are similar to the rules for an aircrew in flight:

- ▲ speak up when you have information that matters to the mission
- ▲ listen up to incoming information that may affect you
- ▲ stand back when others have to do their jobs
- ▲ lend a hand when it's time to act together
- ▲ stay alert

But here's how mission meetings go amiss when that commonsense set of rules is not understood by all participants:

The Derailer

This participant speaks up, but only to derail the train of thought. He begins with phrases such as "What I haven't heard anyone mention is . . ." and "First, I think we ought to consider . . ." and "You all seem to have forgotten that. . . ." What follows is usually a digression into the dusty files of the past or the speaker's favorite anecdotes.

A word or phrase in the derailer's comments usually manages, by association, to set off another derailer in the group. Left to follow its own wandering path, this mode of discussion will lead the group far afield and far from its mission. The most common—and worst—solution is simply to meet again.

Better by far for the chair (or alert participants) to force the derailer to get to his point, if any, and show its importance to the topic at hand. This challenge can be firm but polite: "Wait a minute, George. We're talking about how the FAA evaluates inspections of electrical wiring. What are you saying about that?" If George isn't up to the challenge and continues with his digression, the chair or meeting participants have to cut in again and get the meeting back to its point.

More than one chair, by the way, has lost patience with derailers. Terry Moynihan, Imperial Oil's manager of administrative services, advises that "sometimes you have to bloody well tell someone, 'Keep quiet and let someone else get a word in here.'"

The Instant Replay

This participant missed it the first time and wants to go back, preferably to the beginning. You can recognize him or her by such pleas as "What was your first point?" and "I didn't catch what you said about . . ." and "Could you repeat your explanation of . . . ."

The Instant Replay participant puts both the chair and other participants in an awkward position. On the one hand, it's hard not to oblige the apparently sincere request of a co-worker for a repeat performance of your immortal phrasing. On the other hand—the hand that counts—it's disastrous for the mission of the meeting to get stalled in perpetual repeats.

Turn down the requests of Instant Replay participants by reassuring them: "Let me go on and I'm sure you'll see how . . .," "That will be clear when we get to . . .," or "We have to move ahead now, but after the meeting I'll . . . ."

Don't feel high-handed in such replies. Instant Replay participants aren't always (or even usually) the earnest seekers-of-truth they hold themselves out to be. In corporate life, many of them have learned to get attention in meetings not by making their own points but instead by attaching themselves to the positions of others. Once they've gotten you to repeat—and repeat—your point, it suddenly becomes "our" position for the Instant Replayer instead of yours alone.

The Obstacle

This participant just won't get out of the way. There are fifteen minutes left in a tight meeting, and the heads of three divisions have to forge a compromise. Nothing personal, but the Obstacle just doesn't belong as a speaker in this portion of the meeting.

But there he is, hand poking the angels for the chair's attention. His phrases are attention-seeking and confrontive. "No, no, no . . .," "Wait. Everyone just wait . . .," or "Point of order . . . ." Ignore him. Or, says Goldie Brass of Shoppers Drug Mart, seat him right next to the chair. "You're not going to have him right in front of you waving his hand. It's a way of ignoring him, but not ignoring him, because he has the feeling that you need him by your side to help you in some way."

The Loner

Remember Serpico, the Al Pacino character who stood alone against police corruption no matter what the personal cost? There's one in almost every mid-sized meeting. He will take a contrary position from the group not because the group is wrong or even because the Loner believes in his position, but simply because "Serpico lives." For whatever reason, some individuals resent consensus in any form. For them, all business decisions (and life decisions, for that matter) should be a matter of "yes, but . . . ." They were put on this earth to provide the "but."

This is said not to disparage honest, thoughtful dissent, but to identify a perpetual, habitual dissenter who protests no matter what the issue.

Recognize a Loner in your meetings first by nonverbal signs—a disgusted scowl, a dyspeptic glare, a slouched posture. Then, as if from the bowels of indignation itself, come phrases such as "Not all of us agree . . .," and "You're wrong . . . .,"and "Don't count me in . . . ." Meetings traditionally stop to deal with the vagaries of the Loner. Participants put aside the mission to hear out the often-long diatribe of the one against the many.

In mission meetings, the chair must often move quickly past the pro forma dissent of a Loner who wants his or her heroic moments on stage confronting the dragon of uniformity. The chair may cut in with the terse response, "Richard is not in favor of the motion. Next?" The Loner will probably approach such a chair later to register his complaint about being cut off. This moment provides an occasion to enlighten the Loner about his patterned behavior in meetings and its potential effect upon the mission at hand.

The Specialist

This type of participant contributes in his or her area of expertise, then mentally checks out until his alarm rings—usually one of his key words such as "data base" or "computer memory." Some specialists like to advertise their mental vacations in a meeting by opening a book, writing letters, or doodling. The barely hidden message for others is, "I am so-o-o-o good at what I do that I'm here only as an expert witness."

Specialists won't contribute to the broader development of the mission—"That's not my field"—or to questions that require synthesis—"I'm a computer scientist, not an organizational theorist."

One way to wake up Specialists is to ask them to submit their views in writing, to be delivered by a more active participant. Specialists have been known to stay quite alert in meetings and even to contribute to matters "beyond their field" when the alternative is to be cut out of the action entirely.

Reflecting on his years as president at Intel Corp., Andrew S. Grove points out that "the key to success [in mission meetings] is what the chairman does. Often no one bears that title officially, but by whatever name, one person usually has more at stake than anyone else in what the meeting decides. In fact, the person who calls the meeting is usually the de facto chairman. All too often he shows up acting as if he were just another attendee, and hopes that things will develop as he wants. That tends to ensure that the meeting will be ineffective."

Mission meetings that lack a strong chair provide a field day for the Derailer, the Instant Replay, the Obstacle, the Loner, and the Specialist.

A recent issue of *Supervision* provides a five-question checklist you can use in preparation for a mission meeting:

1. Is it a real problem? Suppose a meeting is called to discuss the problem of the number of trade union grievances being processed in your area. Is this known to be a real problem? Can any conclusions be drawn about the volume of union complaints? If so, what can be done about it anyway? Is it not possible that the real problem exists in another functional area—supervisory training in the handling of people, for example?

2. Is it urgent? A meeting may be called to discuss a shortage of materials. Why now? Is this an urgent situation which has suddenly developed or is it a longstanding "disease" that might be better solved by long-term training of the purchasing staff?

3. Is the true nature of the problem known? Are the facts right? Are sufficient facts known about the situation to be able to identify in it a real, stated problem which will be solved by a group?

4. Is it specific? Is there a specific problem—or is it simply a question of "What is our problem?" Can terms be closely confined in order to discuss one thing only? If people talk about "everything," they will eventually talk about "nothing."

5. Has the group most competent to discuss the problem been invited and is each participant concerned about arriving at a decision? The five minutes it takes a meeting leader to think through these questions will prevent the ultimate business embarrassment: calling an emergency meeting only to discover in the meeting that there's no emergency.

THE PROCESS MEETING

The other half of American business meetings concern themselves more with the process of managing information and people than the product of management in the form of specific action or decisions. Process meetings include:

The information meeting: Here information can flow down from management, up from work units, and across from unit to unit. Presentations are common at such meetings, as are question-and-answer sessions.

The team-building meeting: In these meetings, content takes a back seat to conduct. Meeting leaders and participants demonstrate by example, in buzz groups and panels, their strong desire to work as a team.

The transition meeting: During times of change, the culture of a company undergoes stress and even crisis. Meetings can bring together both those causing the change and those affected by it.

The orientation meeting: These meetings aren't simply for new employees. Many regular staff meetings can be considered orientation meetings: there employees learn who's who and what's what. Status or responsibility levels are assigned, changed, and reinforced. Praise and blame are distributed.

Acting like a mission meeting participant while in a process meeting is a bit like shouting in church. No one will arrest you for it, but you'll certainly feel out of place. Shouts, of course, come in many forms. (Who was it that said, "A yawn is a silent scream"?) Listen for the following "shouts" in your process meetings.

1. "Let's get to the point:" Not a bad idea, except when the point is team-building or information-sharing or any of the other goals of process meetings. Driven, goal-oriented individuals often have difficulty enduring the kinder and gentler world of process meetings. They want something to "happen," though they often fail to consider that positive changes inside people certainly constitute a "happening."

The chair cannot usually pause in the middle of a process meeting to make the rationale of the meeting clear to a "let's-get-to-the-point" type. It just isn't possible, for example, to say, "Jack, we're sharing opinions during this hour of discussion as a way of strengthening cooperative spirit in the company and getting Todd, Alice, and Mark to come out of their shells." Before or after the meeting, however, the chair can make clear to Jack that process meetings differ from mission meetings. If he wants to lead the charge, it would be well to make sure the troops are with him—and that's the work of process meetings.

Division Commander Tom Milldebrandt of the Arizona Department of Public Safety plans for regular process meetings: "I found that one of the principle reasons for monthly division meetings was to give district commanders, who were often separated by miles of highway, a chance to 'vent'—to get frustrations out where they and everyone else could see, feel, and deal with them."

2. "Couldn't this all be done in a memo?" Absolutely not. In many business situations, the medium is the message, in McLuhan's famous phrase. Or, as he preferred to put it, the medium is the *massage.* The medium of the meeting—its process of social interaction—is not the "way" the content of the meeting is delivered so much as it is, in fact, a major portion of that content. Participants get a message from a memo, but an attitudinal and interpersonal massage from a process meeting. Memos provide information to know, but process meetings encourage the willingness to know and to act.

3. "Why tell me? Tell it to the boss!" Some meeting participants don't understand why information flows *out* before it flows *up*. For example, a mission-oriented line supervisor might not want to spend time discussing a new benefits plan. "It's not my decision. If you have an idea, send it upstairs." Process meetings provide an environment in which ideas are "tried out" and support is sought before upward communication begins. The answer, then, to "Why tell me?" is straightforward: "Because together our voices are twice as loud."

Sales meetings provide good examples of the process approach. Especially when tens or hundreds of sales people are brought together, there's rarely a single issue to be debated or "mission" to be accomplished. Instead, says Richard Day, senior vice president of group insurance at Bankers Life, "I try to give the newer salespeople and those who are somewhat less successful the idea that they are playing with winners, that they have a chance to grow with the winners." He plans meetings that are "a kind of cross-pollinization and synergism that keeps everybody high."

At Biomaterials International, Director of Sales and Marketing J. Michael Brown has a similar approach. "Sales meetings should be used as a reward. Take your people to a place that's relatively exotic—as much as you can afford—and don't make them spend the entire time working their fannies off in a hotel!"

In sum, process meetings try to achieve internal goals within individuals and groups. Mission meetings try to achieve external goals. Of course, real business life often forces these two types together into a single meeting. A staff meeting aimed at team building, for example, may also have to deal with a pressing problem.

Knowing when to speak and how to speak in a meeting switching from process to mission and back again is difficult and exciting. One CEO likened the challenge to jumping aboard a moving train. Jump too soon and you're flattened. Jump too late and you'll land on the tracks.

THINGS OR PEOPLE?

Whether conducting or attending a mission meeting, a process meeting, or a mixture of the two, you will have to decide how to communicate best with your audience. Tom McMahon, director of Meeting Expectations, a Toronto consulting firm, tries to determine if a meeting audience is "thing-oriented" or "people-oriented." For McMahon, thing-oriented groups "are comfortable dealing with facts, figures, data, logic. They like to see slides up on the board that show percentages and charts. They like taking notes and doing their thinking privately."

You Probably Shouldn't Call a Meeting If . . .

- you have no clear agenda in mind
- you're relying on meeting participants to come up with agenda items [Call participants well in advance of the meeting, then organize their suggested agenda items into a logical agenda.]
- you know you need to meet with one or more of the meeting participants, but don't know which [Make a few phone calls and spare the group an unnecessary meeting!]
- you could postpone the meeting without causing problems
- you plan to do all the talking in the meeting [Put your thoughts down in writing, perhaps in the form of an extended memo. Ask recipients to call you if they have questions or concerns.]
- you have already made your decision and want to convene the group merely as a rubber stamp [Again, use a memo to communicate decisions and policies.]
- you want to get together simply to get together [Have a party, not a meeting!]
- you need a meeting simply to show a superior you're doing something [A bad meeting can't make a good impression.]
- you want to spend the meeting getting answers to your questions from individual participants [Call the individuals.]
- you have no idea how long it will take to deal with your agenda items [Don't find out in the meeting itself.]

People-oriented groups, by contrast, may be bored by the same slides that fascinated a thing-oriented audience. Says McMahon, "people-oriented people are interactive. They like to learn through people" in the form of informal presentations, question-and-answer sessions, buzz groups, and ordinary conversation.

The great majority of meeting participants—some 80 percent, according to Daniel Stoffmann in *Canadian Business*—are people-oriented. Yet, says Stoffmann, "most meetings are designed for thing-oriented people." The balance can be readjusted by including group participation activities in meetings. Bryson Chapman, president of Communique Ltd., likes to use a multiple-choice quiz for a people-oriented change of pace. Each meeting participant uses a hand-held device much like a channel selector to indicate possible answers to questions that appear on the screen. In this way, audiences can be asked questions ranging from opinion surveys to retention quizzes on a speech just given. "It encourages people to pay attention," says Chapman.

Another people-oriented technique used frequently by McMahon in his consulting practice is role-playing, or "Put yourself in the customer's place." McMahon chooses one participant to play the part of a customer and another for the role of a company representative. The group watches the pair's improvised skit, then discusses pros, cons, and alternatives. "A game like that is a lot of fun," McMahon says, "but it also leads somewhere."

A WINDOW ON REAL MEETINGS

We'll look in now on meetings in three industries. Compare the opinions expressed here with your own experience in these fields.

Meetings in the Entertainment Industry

Take one. Take two. Take a dozen. The fact is that Hollywood has far too many meetings for its own good. David Brown, producer of *The Verdict*, *Jaws*, and *The Sting*, says that "in the entertainment business today people are unable to return their phone calls. Why? They are in a meeting. They are always in a meeting. More meetings are held and more business is lost because you can't reach people."

Many of these are "power meetings" called simply to give high visibility to upcoming projects or studio developments.

Typically these meetings are larger than they should be, populated by PTIs—People To Impress.

The way out of too many meetings in the entertainment field, says Brown, is to recognize that "less is more in a meeting. Less people, less time. American business would be much better off if meetings were contained to 30 minutes at 7:30 in the morning." Many of Brown's own meetings now take place over breakfast.

Meetings in Informational Processing and Financial Data

If employees were connected by electronic mail, computer bulletin boards, and other hi-tech communication technologies, would meetings become unnecessary? Not at Dana Corporation, as described by researcher Steven A. Stromp. Manager Ron Parks of Dana Corporation can communicate by computer with any of the 51 employees. Yet, says Stromp, "every Monday at 9 A.M. Parks goes to the plant with the plant manager to meet face-to-face over coffee with the workers. The weekly sessions are Dana's way of letting all of its employees know the score—whether the company is winning, losing, or breaking even with the competition."

Such meetings now take up a large part of each manager's day at Dana Corporation. Parks explains that "we try to meet with the people once a week for 30 minutes to an hour to let them know what's going on and what's our status." The door is always open to questions. "Employees often ask why a decision is made. And they get an answer. If they aren't satisfied with what we tell them, they can direct questions up the ladder—right up to the chairman."

In this way, companies like Dana Corporation are fulfilling John Naisbett's prediction that "hi-tech" will have to be accompanied by "hi-touch" for ultimate business success. The labor-intensive process of meeting real people face-to-face, warts and all, isn't as easy as sending computer messages. "Good communications are not easily done," says Parks. "It's a chore every time. But we at Dana make an effort at it because it's important."

Meetings in Sales Organizations

In a 1985 survey of companies with large sales forces, researchers asked what kind of sales meeting each company planned to hold in the coming year. The survey team was surprised by one answer, from a midwestern leather products manufacturer: "We have no reason to hold one [a sales meeting]. Sales are good!"

The same survey revealed interesting facts about American sales meetings:

- ▲ Almost half of the companies surveyed budget between $10,000 and $50,000 for an annual sales meeting, excluding travel. More than one third of the companies budget less than $10,000.
- ▲ The responsibility of choosing both the meeting site and the meeting agenda goes most often to the vice president of sales/marketing.
- ▲ More than 90 percent of companies hold regional sales meetings with a total attendance of 50 or less people. More than 80 percent of companies hold national sales meetings with a total attendance of 100 or less people.
- ▲ Two-thirds of the respondents indicated that they were trying to decrease rather than increase spending on sales meetings.
- ▲ The number of companies spending $100,000 or more is steadily dropping from previous years.
- ▲ Companies are turning away from resorts in favor of airport hotels, company headquarters, and budget motels. Researcher Leslie Brennan, writing in *Sales and Marketing Management,* reports that "resort hotels suffered the biggest drop of all sales meetings sites in 1985, declining five percentage points from last year's 42 percent. Company sites and airport hotels picked up the difference."

The continuing emphasis of regional and national sales meetings is less educational than inspirational. As process meetings, these gatherings should deliver what Richard P.

Day of Bankers Life calls "a kick"—a motivational boost stemming from a few days of special treatment, praise, and camaraderie.

At the other end of the meetings spectrum is the small sales meeting, often involving only a salesperson and clients. These meetings are an even balance of process (nurturing a relationship with the client) and mission (making the sale).

Effective sales meetings take place at the salesperson's office, the client's office, a conference, or restaurant. As the following chart shows, they are best scheduled in the morning, and at selected times in the afternoon. Avoid the "dead" time immediately after lunch.

Finding the Best Time for Sales Meetings

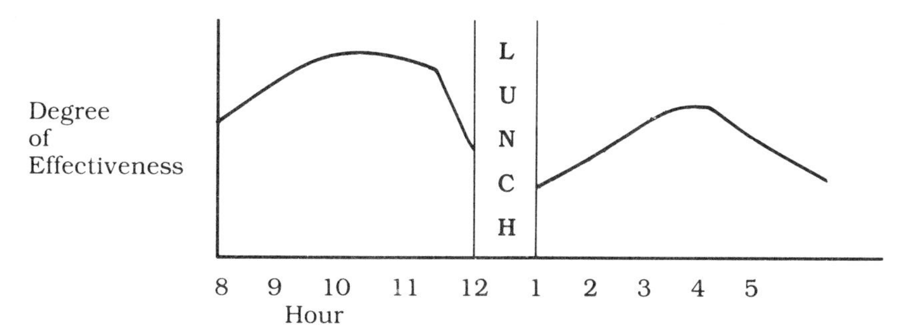

The typical progress of the sales meeting with a client moves according to the "5 P's": People, Purpose, Probes, Presentation, and Production.

"People"—Establish good interpersonal relations by your pleasant demeanor, sincere interest, and friendly conversation.

"Purpose"—Move without unnecessary delay to the reason for your meeting. State it clearly, using whatever input the client offers.

"Probes"—Ask probing questions to determine the client's specific needs.

"Presentation"—Deliver your sales presentation in an energetic, natural way. Don't give the impression you've memorized your words.

"Production"—End the sales meeting by *producing* something: a signed sales contract, an agreement for a future meeting, or some other form of commitment from the client.

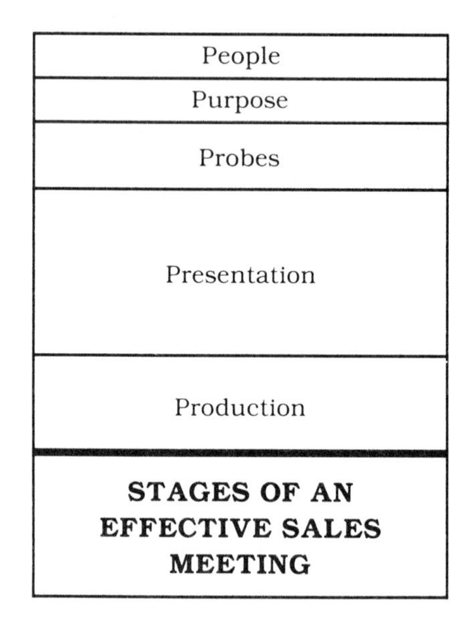

STAGES OF AN EFFECTIVE SALES MEETING

One important key to effective sales meetings, according to Kate Reilly and Eric Baron in *Business Marketing*, is planning "the how part of sales calls. No one likes canned presentations or rehearsed sales pitches, so salespeople should not go in with a script."

ROLODEX RECAP

▲ To participate effectively, meeting attendees must understand the kind of meeting they're attending.
▲ The mission meeting focuses on decision-making, problem solving, and policy formulation.
▲ Mission meetings are often disrupted by the behaviors of the Derailer, the Instant Replay, the Obstacle, the Loner, and the Specialist.

ROLODEX RECAP

▲ The process meeting focuses on information sharing, team building, transition adjustment, and orientation

▲ Process meetings are often disrupted by those who act as if they're in a mission meeting

▲ How you communicate in a meeting will depend on whether your group is "thing-oriented" or "people-oriented"

Finding Out What People in Your Organization Think about Meetings

Distribute this questionnaire to a significant cross-section of your workforce.

1. Approximately how many hours each week do you spend in meetings? ______
2. What kinds of meetings do you typically attend?

   ____

   ____

   ____

3. For each of these meeting types, choose a number that expresses your overall evaluation. Place the number in the blank beside the meetings you named in #2.
   1—always productive and worthwhile
   2—usually productive and worthwhile
   3—occasionally productive and worthwhile
   4—seldom productive and worthwhile
   5—never productive and worthwhile
4. Circle the phrase that best expresses your opinion about the total amount of time per week you spend in meetings:
   too much time   *about the right amount of time*
   too little time
5. Circle the phrase that best expresses your opinion about the length of meetings you typically attend:
   too long   *about the right length*   *too short*
6. Circle as many of the following phrases as you would use in describing this company's meetings:

tightly organized	*good use of time*
dominated by the meeting leader	*disorganized*
	valued by particiupants
loosely organized	*disliked by participants*
little discussion	*action-oriented*
poor use of time	*too little follow-up*

7. For more productive meetings, what one suggestion above all others would you give to meeting leaders in the company?

____

____

____

____

8. For more productive meetings, what one suggestion above all others would you give to meeting participants in the company?

____

____

____

____

3

Leading a Meeting

I MET in New York this past January with a prominent industrialist—a man known for leading lively and efficient meetings. He remembers other days:

> "Early in my executive career, I took over the helm of a struggling company. I hand-picked an executive council made up of some of the company's seasoned managers and called them together for our first meeting. Tuesday, 10 a.m.—I remember it too well.
>
> I began with introductory remarks, then asked for questions. There were none. I took a few minutes to explain the first item on the agenda, then asked for discussion. There was none. I talked for a while about the topic and finally just called for a vote. This went on for each agenda item. I couldn't get them to talk!
>
> For the next meeting, I assigned several reports to be given. They were dutifully read, I gave my reaction, and then silence fell again.
>
> I felt like I was at a funeral. Had I picked the wrong people? Was it me?"

This executive was experiencing the *Head Honcho Effect.* And no wonder. In traditional organizations, the man or woman at the top carries an atmosphere, an influence, into a meeting or conference. Often these leaders are the last to recognize the cause of their problem. So let's spell it out.

THE OLD VIEW OF THE MEETING LEADER

In preparing for this book, I asked fifty supervisors and mid-level managers about their attitudes toward meeting leaders. The questionnaire asked a straightforward question: "How do you feel as a participant when an executive in your company leads a meeting?"

Consider some of the most common responses:

> "Somewhat threatened—like anything I say may come back to haunt me. So I usually don't say much."

> "I often feel that the boss who's leading the meeting knows most of what we have to say anyway. That's what being the boss means."

> "It's a ritual. The executive in charge usually has a predetermined game plan in mind."

> "I feel like a spectator. People like me in the company trenches don't get to see the boss in action very often. It's a chance to sit back and watch."

The boss as Threat, All-knower, Game-player, or Main Attraction—these are all clear symptoms of the Head Honcho Effect.

And death to effective meetings.

Participants who feel threatened by the meeting leader clam up. Those for whom the leader seems all-knowing don't make an effort to contribute their own insights. Those who think of the meeting as a predetermined game feel and act left out. Finally, those who view the meeting leader as a flash from Las Vegas simply sit back and watch. The result: silent, stagnant, boring meetings.

LEADING MEETINGS IN THE 1990S

To unlearn the habits that lead to the Head Honcho Effect, we can look to Japanese "Theory Z" leadership techniques. In the famed "quality circles" used throughout Japanese business and industry, top executives meet with a wide

range of their employees. These meeting leaders make four meeting guidelines absolutely clear:

1. *Nothing you say in this meeting will be held against you.*
2. *No one here has all, or even most of, the answers.*
3. *No one, especially the meeting leader, will be spotlighted as a "star" in the meeting.*
4. *No one has predetermined what should be said or decided in the meeting.*

In a word, the meeting leader tries to be invisible—"the catalyst agent," says a Japanese manager, "that makes other ingredients combine."

Certainly American business meetings do not and should

not mimic all Japanese business customs. Our cultures are too different. But American meeting leaders can borrow creatively from the Japanese notion of the "invisible leader." And, says Tom McMahon, president of Meeting Expectations, Inc., don't worry about losing "control," which is "just the opposite of domination. Control means acting so professionally and so well as a facilitator that you're not even noticed. The meeting seems to have a life of its own. It progresses without anyone noticing that they're being kicked in the pants or dragged into the next phase."

It's an ineluctable fact that meetings cannot *not* be led. In the absence of leadership from the chair, members of the group will take charge. Contractor Richard Steinke recalls attending a large sales meeting in which the chair allowed discussion to drift overtime and far from the agenda. "In absolute frustration," says Steinke, "I started to clap. Then the whole group started to clap. The chair got the point and ran the rest of the meeting on schedule."

Here, step by step, is a "walk-through" of a typical American business meeting through the eyes of its leader—a corporate vice president determined to avoid the Head Honcho Effect.

THE INVISIBLE LEADER IN ACTION

Shrewd meeting leadership begins before the fall of the gavel (or, more often, the clink on the water glass or loud "Ahem . . ."). In a natural, comfortable way, the meeting leader has pressed a few palms, patted a couple of backs, and passed pleasantries with most meeting participants as they entered the meeting room. This openness goes far to relieving the sense of threat that many employees, especially new hires, often feel.

We'll assume that the meeting leader has provided each participant with an agenda well in advance of the meeting (see Chapter 5 for a full discussion of agendas). For now, we'll focus on the leader's verbal and nonverbal signals.

Do you recall walking into a meeting and seeing the leader

in close, guarded conversation with a few intimates? How did you feel? A bit on the outside looking in? Leaders have to break the ice early in making each meeting participant feel welcome to the session. (Don't misunderstand: a Miss Manners hostess or host you don't have to be. Just make sure meeting participants don't form a first impression of you as The Shadow.)

Our meeting leader keeps her opening remarks very short—she doesn't want to be "on stage" with everyone else sinking into their chairs as spectators. After a cordial welcome and clear statement regarding the purpose of the meeting, the leader calls for brief committee reports.

This handoff works well for two reasons. First, it signals to the group that the leader welcomes input from others. Second, it gets members talking. Interesting research at Georgetown University School of Business Administration demonstrates that once a person has spoken up (even briefly) in a meeting, he or she is much more likely to speak a second or third time. Those who do not break the ice early in a meeting may tend to play Silent Sam for the entire session.

Encouraging Participation Every manager knows the truth of the old adage, "The carrot works better than the stick." Praising meeting participants for their contributions does work, but only when such reinforcement is sincere. And there's the rub.

"I felt stupid," admits a Connecticut bank president, "passing out inane reinforcers like 'Thanks, Al, for that comment' and 'Good point, Alice.' I felt like some kind of student teacher. And my comments didn't seem to work. They sounded phony."

Susan Todd at a Silicon Valley computer company agrees. "In leading my first meeting of systems designers, I knew I was supposed to encourage people to give input. But I felt like a judge passing out verdicts: 'Yes, Barbara, that's right,' and 'I agree with you, Frank.' The focus kept coming back to *me* and what *I* thought."

"Now this is called 'getting off your duff.'"

Try two techniques to avoid treacly, insincere reinforcement:

1. Let other members of the group give feedback. When someone speaks up, he or she usually isn't looking for a pat on the head or a swat on the behind from the meeting leader. In most group dynamics, each individual is seeking the support of the surrounding social group. Allow that support to be given (or denied) by passing the ball to another group member. Assuming Frank has just spoken up, the meeting leader says: "Jill, you've worked with that software. Do you agree with Frank?"

You've given Frank encouragement in a sincere and nonthreatening way: you've listened to him and asked the group to consider his viewpoint. Notice that you haven't passed judgment on him. Let the group do that. (To some degree you can influence the amount of praise a contributor receives by whom you call on. In this case, you knew that Susan agreed with Frank. As an "invisible" leader you are shaping the beginning of consensus around Frank's point of view.)

2. Educate your people on meeting skills—in effect, a series of meetings on meetings. As Georgetown Professor of Management Dayle Smith says, "most of us know only bad meetings. We've sat through them for years. We only know how to participate in bad meetings."

Train for new meeting attitudes and habits in your organization by discussing what works and doesn't work; by passing out good books or articles on meetings; and, by showing a savvy film or two. (The Monty Python group stars in two stellar flicks mentioned in Chapter 1: "Meetings Bloody Meetings" and "More Bloody Meetings." Dozens of other films and videotapes are listed at the end of Chapter 9.)

Active listening Back at the ranch, our meeting leader is looking eye-to-eye with Wilma, speaking up from the back row. Wilma usually has a bad case of the "shy's" at meetings—but this time she's expressing herself fully and energetically.

What made the difference? *Active listening* on the part of the meeting leader. By her steady eye contact, her nods of interest (not necessarily agreement), and her slight lean forward to catch each word, the meeting leader shows Wilma that her opinion matters. These are the nonverbal demonstrations of interest that even famous corporate leaders often neglect.

At one time or another, we've all been "called upon" by a meeting leader, who then looks away to notes or to other group members. Like a Senator speaking to an empty hall, we feel that we're speaking into the void. Only when we've finished does the leader seem to spring back to life. The group's impression of the event isn't rosy: "The chairperson looked bored with that comment. It must not be important."

Active listening for the meeting leader can involve more

than your eyes, facial expression, and posture. Notice in the following script how the meeting leader encourages full communication by "saying back" key phrases from a group member's contribution:

MEMBER: "I agree that we need a new accounting method for Accounts Receivable, but it has to include some flexible way of using outside collection agencies . . . to, uh . . ."

LEADER (*shows interest, encourages*): ". . . collection agencies"

MEMBER: "Yeah, we use them every year, whether we plan to or not. There are always certain accounts that just don't pay. We may as well get even a small percentage out of them instead of nothing."

LEADER (*says back*): "You're saying that a new accounting method should include categories for outside collection."

The technique of "say back" (pioneered by psychologist Carl Rogers) may seem obvious and trite on the printed page—but works like magic in the heat of discussion. We all like to feel that others "really listened." The group, at the same time, appreciates the highlighting and review of key points and phrases.

The technique of "say back" can easily be overdone, of course. Like insincere compliments, this form of active listening has to be natural and comfortable for all parties to the communication. A good way to observe the technique is to notice, in your next extended conversation with a close friend, how often your own words or expressions of emotion are reflected back to you by the friend. At these moments in the conversation, you may find yourself saying or feeling "that's right, that's exactly what I mean." You're responding, of course, to your own words. But when they come back at you through the mouth and expressions of your friend, they seem to be a confirmation of what you're trying to say.

"We brought in a high-priced consultant to teach us the virtues of active listening in meetings," reports Ellen Gillson, a senior manager at a Florida resort. "It worked fine in the seminar, but flopped in actual meetings. We couldn't listen actively because of the way we were sitting! Our meetings were held classroom-style, with everyone looking up front to the meeting leader."

The Communication Revolution, with its beepers, portable phones, and pagers, poses new challenges for focused, attentive listening in meeting. Often the group has to agree to banish the technology during the meeting. Police Division Commander Tom Milldebrandt insists that at his meetings "all cellular phones, pagers, and walky-talkies are left outside. A staff member screens all calls and all but real emergencies are handled on breaks."

If your meeting room doesn't allow faces to look directly at other faces, change the room. (Plush executive "meeting theatres" are often disastrous for active listening. The rooms have been designed for eyes-forward viewing of slide presentations and executive speeches.)

Most of the seating plans described in Chapter 5 will guarantee that each participant can give active listening signals to anyone who speaks.

Keeping an Eye on the Clock Our meeting leader is doing just fine, thank you. She's a supportive presence who encourages participation and listens actively. What more could meeting participants want from her?

That question, in essence, appeared on a 1987 questionnaire completed by more than 500 workers across industries. The answer was surprising. One of the three qualities employees desired most in a meeting leader was "getting us out on time." Joan Hamann of the University of Wisconsin's Management Institute encourages a meeting leader "to be a ruthless clock-watcher who keeps the meeting within its prescribed time limits. . . ."

Employees at all corporate levels, it turns out, give high marks to any leader who can end a meeting on time. The group may cherish its freedom to discuss, but not its tendency to gab on and on. Group members want the leader to have a firm rein on how the group uses its time.

At key points in the meeting, a leader can remind the group of the clock. "We've spent ten minutes discussing the pros and cons. Do you want to move to a vote, or shall we continue discussion?" Or, "we have about 20 minutes left before our agreed cut-off time of 5 p.m. Let's each limit our final remarks to no more than two minutes each."

Meetings that drag on can tempt the group to act irresponsibly. School Board President Janis K. Long recalls overly long meetings she's attended in which "the group became fretful and willing to agree on *anything* just to get on with it."

The chair's occasional reminders about time are appreciated by the group—especially when members know they can speak up to extend the time limit if necessary. Most of the time it isn't necessary—very, very few business people take up the cry, "No, let's meet even longer!"

For long meetings—"which I define as one that lasts more than one hour," says Cost Administrator Keith Conway at Ford Aerospace—participants can be divided according to when they're needed. In Conway's meetings, "the core group may be needed for the whole meeting while the specialists are scheduled in for particular agenda items."

As shown in the Action Agenda in the Appendix, suggested time limits can be placed next to agenda items. So long as they are not enforced too strictly, these limits will be accepted and even appreciated by the group. With practice, meeting members will learn to shape their comments and debate to fit within the time period available. Their reward, of course, is finishing on time—which often translates into a hot dinner instead of a cold one.

Sometimes it takes administrative guts to adjourn early. At Loral Defense Systems, Program Director Ralph Franke remembers the "worst meeting I ever attended. It was a brainstorming session where the problem at hand was solved in very little time. Unfortunately, people had come long distances and were not ready to accept the obvious and disband. Instead of adjourning the meeting, the leader let it drag on all day. An hour or two probably would have been adequate."

Clarifying Issues Our meeting leader now has encouraged participation, rearranged the room for active listening, and kept the group on track and on time. In effect, she's made each participant feel like a quarterback, not a couch potato.

But a field full of quarterbacks poses its own problems. Plays come fast and furious, and get confused. So many opinions are put forth by animated members that the group feels paralyzed by alternatives. "Shell-shocked," says Dale Larson, vice president of Norwest Bank. "Meeting members with overloaded circuits are unable and unwilling to absorb any more information."

Enter the meeting leader to clarify issues. Here are some useful openers for clarifying remarks.

"Let me see if I understand what several of you are saying. . . (*proceeds to highlight key issues*)"

"Before we move on to that new point, let me try to sum up where we've been . . ." (*summary follows, with a consensus check—"Am I on track so far?"*)

"So you're saying we have to make one of three choices . . ."

"Just to make sure everyone is clear on that point, let me say it back to you in my words . . ."

"What I hear you saying, Jack, is . . ."

Our meeting leader doesn't clarify or summarize unnecessarily. She watches the group for clues that such comments will be welcome. When group members listen with sustained frowns, flash quizzical or frustrated looks at one another, begin private conversations, shuffle papers, and so forth, the meeting leader intervenes with a brief clarifying or summarizing remark.

"But don't let explanations and summaries sound like something from elementary school," says Nurse Betty Sherick at Good Samaritan Medical Center. "Often we get the point without the 'in other words' clarification from the meeting leader."

Using a greaseboard or flip chart can help a group focus on issues and ideas. For major portions of the meeting, the leader's role may simply be that of recording secretary, jotting down key points on the board—'immortalizing contributions,' in the words of Kevin Daley, president of Communispond, Inc. This simple act can encourage members to contribute, since they know their ideas are being taken with some seriousness, and can steer a course away from confusion. No group, after all, should realistically be expected to hold in mind four, five, or six alternatives without help from a flip chart or greaseboard.

The meeting leader, busy writing these notes, doesn't have to take on the burden of providing clear summations of each issue for the group. Notice how the following handoff allows the meeting leader to focus discussion while avoiding speeches:

"I've noted each of your suggestions on the flip chart. Does someone want to recap the first one before we begin discussion?"

Lack of attention, says American Express vice president Robert Munson, can be handled effectively "by calling attention to the parties involved." Or the chair can physically move closer to parties who are carrying on private conversations.

Soliciting Feedback An early notion of corporate communication pictured words as tiny dumptrucks. You fill your dumptrucks with ideas, send them over to my head, and—barummph—your ideas get dumped into my head. Voila, communication!

But our meeting leader knows that this one-way view of communication is flawed. Communication in meetings can't be said to be complete until the message sender receives *feedback* from the message receiver. (Similarly, a car salesman can't claim, "I sold the car, but they didn't buy it." The act of selling includes the act of buying, just as the act of communicating includes the act of feedback.)

We've already pointed out that the meeting leader shouldn't play judge ("Good point, Frank!") for each contribution from group members. How does a member get feedback?

First, the faces of other group members scream out silent messages: "We agree! We disagree! We don't understand! We have a question!" Speakers who can read these messages even while speaking have a chance to adjust their words to the needs of the group. (Again, the room must be arranged so group members can see one another.)

Second, the feedback process can be encouraged when a meeting leader simply asks for feedback:

- "What do you think of that perspective?"
- "Janet, how would you evaluate Bill's suggestion?"
- "Those of you from C Sector probably can comment on Mike's point. How do you see it?"

By such prompts, the meeting leader helps the group "close the loop" on ideas and move toward decisions. Nothing is more frustrating than an "idea orgy" where a host of opinions are exposed, none of them leading to the marriage of minds.

Feedback gives the group a chance, step by step, to make up its mind, even in a tentative way, about opinions and viewpoints. Just as important, a bit of feedback gives the speaker a chance to assess the weight and worth of his or her own spiel. After all, much of what's said in a meeting isn't born of heartfelt commitment so much as situational experiment. "Float it out and see if it flies," many group members say to themselves just before speaking out.

Feedback provides an answer—"yes, it flies," "no, it flops," or "well . . . maybe."

Channeling Conflict. Oh-oh. Problems in paradise. Our meeting leader, who so far has guided discussion with a deft hand, now faces an out-and-out argument between Calvin and Ruth.

Our leader accuses herself: "I should have prevented this argument from happening." There's her mistake.

Strong disagreement can often be considered a positive outcome of a meeting. That conflict, if not uncovered in the meeting, will certainly pop up elsewhere—at the water cooler, through the grapevine, or in arcane personnel and procedural battles.

Paul Velasquez, who manages a Texas food-processing plant, faced up to the problem of conflict at meetings. "I can't name names, but suffice it to say we had a major personality duel between two of our key people. At first, I would invite only one of them to a meeting at a time.

"I soon realized, however, that I was just forcing the battlefield out of my sight. They were still having it out, but not in my meetings. So I brought them both to the same meetings. It wasn't pretty, but I can say that in our meetings we've worked through their conflicts in a controlled and useful way. They still don't like each other, but at least they're playing on the same team again."

The meeting leader has to act fast at the first flash of conflict:

1. **Make sure the group focuses on the issue.** The flare of emotions will tend to obscure the content of the issue. Without leadership, the group will automatically focus on "George dislikes Willard," not "George dislikes Willard's position."

Note in the following statement how a meeting leader can focus on content instead of emotion: "OK, OK, let me break in for a second. George, you're making the point that maintenance contracts should be purchased at the same time we buy the computers, not afterward. Is that right?"

Meeting consultant Tom McMahon says that "the best way to stop an argument is to reinterpret the issues in a nonpersonal way. If George, the controller, has just called Sandra, the sales manager, 'an airhead who can't even count to 10,' the chairperson should step in with something like 'What you're saying, George, is that you need much more communication with Sandra to understand her function, and the real issue. . . .' He can then give Sandra and George some time to cool off by immediately asking someone else for his or her comments."

2. Don't take sides, even in your nonverbal nods and gestures. You may want to remind the group that "it's important to hear both sides out, but let's focus on issues, not personalities."

3. Don't rush to a solution. In doing so, a meeting leader only gives the conflicting parties an additional motive for battle. The turmoil can be calmed somewhat by pointing out "we need to air this issue. There's no need to find an absolute answer right here on the spot."

4. Don't call attention to personal attacks between conflicting parties. If at all possible, simply ignore snide and sarcastic remarks. Focus on the content at issue: "Let's not stray from the issue. . . ."

5. If possible, ask each side in the argument to state the other side's point of view. This exercise often helps parties realize that they have more to agree on than they thought.

Our meeting leader, experienced as she is, can often spot potential conflict well before the meeting starts. She influences and partially controls such conflict by her choice of seating arrangements.

By seating opposed parties across from one another, she virtually assures that conflict will be expressed maximally during the meeting:

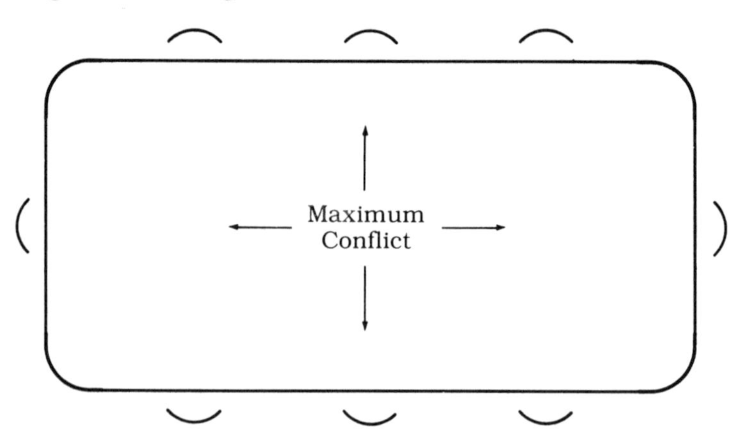

By seating opposed parties next to one another, she may lessen the tendency for open conflict. It's hard to scream at someone sitting right next to you:

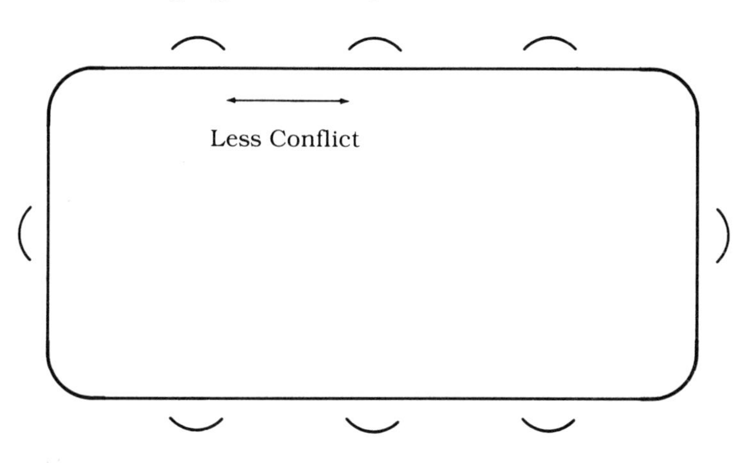

Seating in most meetings can be subtly dictated by the preparation and placement of a meeting folder bearing the participant's name. The great majority of meeting attendees will sit where they find their folder placed.

MANAGING GUEST SPEAKERS

Joanne C. Jeanguenin, executive director of the New Jersey Society of Association Executives, lays the responsibility for what guest speakers say and do right in the lap of the meeting leader: "Imagine that you've contracted the best time-management speaker in the country. He's prepared to talk about the need for personal goal-setting and self-management. But the attendees are expecting nuts-and-bolts tips. The result? A terribly mismatched program and audience." Jeanguenin recommends that a leader "determine what the audience expects from the meeting," then "share that information with the speaker."

University of Massachusetts Professor Mack Drake points out that guest speakers too often "immediately call for lights out and launch into their slides. It is much more helpful when speakers establish audience contact by describing a bit of background or highlights of their presentations."

WHAT ABOUT HUMOR?

Natural humor that doesn't overtake the business purpose of the meeting is certainly appreciated by attendees. Staged, rehearsed jokes usually flop. As Stanley Bing writes in *Esquire*, "you don't have to be very funny to kill a corporate audience. Just the mere intimation of wit often sends a room of hitherto gloomy executives into paroxysms of laughter. I've heard punch lines like 'It certainly is' or 'No kidding, Jack' wrest a torrent of hilarity from a somber room, with senior management removing hankies to wipe their eyes from the surfeit of sheer amusement."

As with most things, the key to effective humor is spontaneity. Paraphrasing Keats, "if humor comes not as naturally as leaves to the tree, then it should not come at all."

READY FOR EMERGENCIES

A recent issue of *Business Marketing* points out that "minor things can spoil an important meeting. The projector cord that doesn't reach the outlet. The blackboard with no chalk. The flip chart with no markers. The microphone with static. The projector bulb that blows out halfway through the meeting. The loose screw on the flip chart easel that causes it to sway precariously."

This litany is familiar to any experienced meeting leader. I recall one very large gathering I chaired in the Seattle Opera House. Midway through the meeting, the audience of five thousand were supposed to enjoy a brief slide show projected onto a huge screen by a powerful arc projector. Instead, the audience was treated to several minutes of pitch-black darkness—the longest minutes of my life. The carbons had burned out in the arc projector.

At such moments, it isn't enough silently to blame a technician. Egg gets splattered all over the meeting leader.

Though the mishap may not be his or her fault, it nonetheless damages the meeting.

Business Marketing recommends a "first-aid kit" for meeting leaders. It costs less than $100 to assemble and may be worth its weight in gold over the course of meetings in your career. The kit should include

- chalk
- eraser
- masking tape
- blank transparencies
- bold felt-tip pen for flip charts
- lavaliere mike (available at electronics stores)
- extension cord
- 110-volt, 15-amp two-prong electrical adapter plug
- pens
- pads
- screwdriver set
- extra bulb for overhead projector

RULES OF ORDER

The great majority of business meetings in America use commonsense rules of order, as interpreted by the chair. Each agenda item is taken in order, with discussion and a vote or demonstration of consensus. Such parliamentary procedures as "calling the question" and "point of order" are not used, or are rephrased by attendees in such suggestions as "let's see if we can reach a decision" or "I want to ask a question about the process we're following."

If your organization or corporate culture makes use of more formal meeting procedures, the standard guide is still *Robert's Rules of Order.*

THE CHAIR'S SUMMARY

As chair, you began the meeting with brief introductory remarks that spelled out the purpose and set the tone for the gathering. It's up to you at the end of the meeting to reinforce that purpose, telling participants to what degree the meeting accomplished its intended goals:

> "I want to thank each of you for making time in your schedule for this emergency meeting. I think we ironed out 90 percent of the administrative problems involved with the Apex contract. Later this week I'll meet with you individually to wrap up the last details. Thanks for your good work."

Whenever possible, the chair tries to send the group out of the meeting on a high note—effort well spent, mission accomplished. He or she follows up on responsibilities delegated during the meeting. Marketing consultant Marlene Rossman, for example, sends around a memo to attendees a few days after the meeting. She gives her version of the highlights of the meeting and the tasks passed out there. Later she recirculates the memo and asks her people to update their progress on assigned responsibilities. In this way, the agenda for the meeting spawns an action agenda—in the form of a memo or minutes—that orients work after the meeting.

Julie Bailey, a former editor of *Business International*, uses a Broadway analogy to sum up the challenge of being a meeting leader: "A successful chairperson plays two roles—director and lead actor. Double duty calls for plenty of preparation, sensitivity, and often a willingness to share center stage with other performers. Learning to carry it off with skill and brio guarantees a reputation for command performances and the kind of enviable results that enhance a manager's reputation and can boost her [or his] career."

There's no greater compliment to a meeting leader than the calls of upper management and meeting participants for you to lead the next meeting: "Encore!"

ROLODEX RECAP

▲ meeting leaders must guard against the 'Head Honcho Effect'—the tendency of the group to think of a leader as Threat, All-Knower, Game-Player, or Main Attraction

ROLODEX RECAP

▲ the 'invisible' meeting leader provides an agenda; makes friends with the group early, keeping opening remarks short; encourages participation; listens actively; watches the time; clarifies; solicits feedback; channels conflict; and follows up

ROLODEX RECAP

▲ The meeting leader works in advance with guest speakers to make sure they know their audience, purpose, and time limits.

▲ The meeting leader uses only natural, unrehearsed humor.

▲ Technical emergencies are prevented by a 'first aid kit' of cords, bulbs, markers, and other items.

ROLODEX RECAP

▲ Most meetings follow commonsense rules of order as agreed to by the group.

▲ The meeting leader sums up the main action of the meeting and relates it to the purpose set for the meeting.

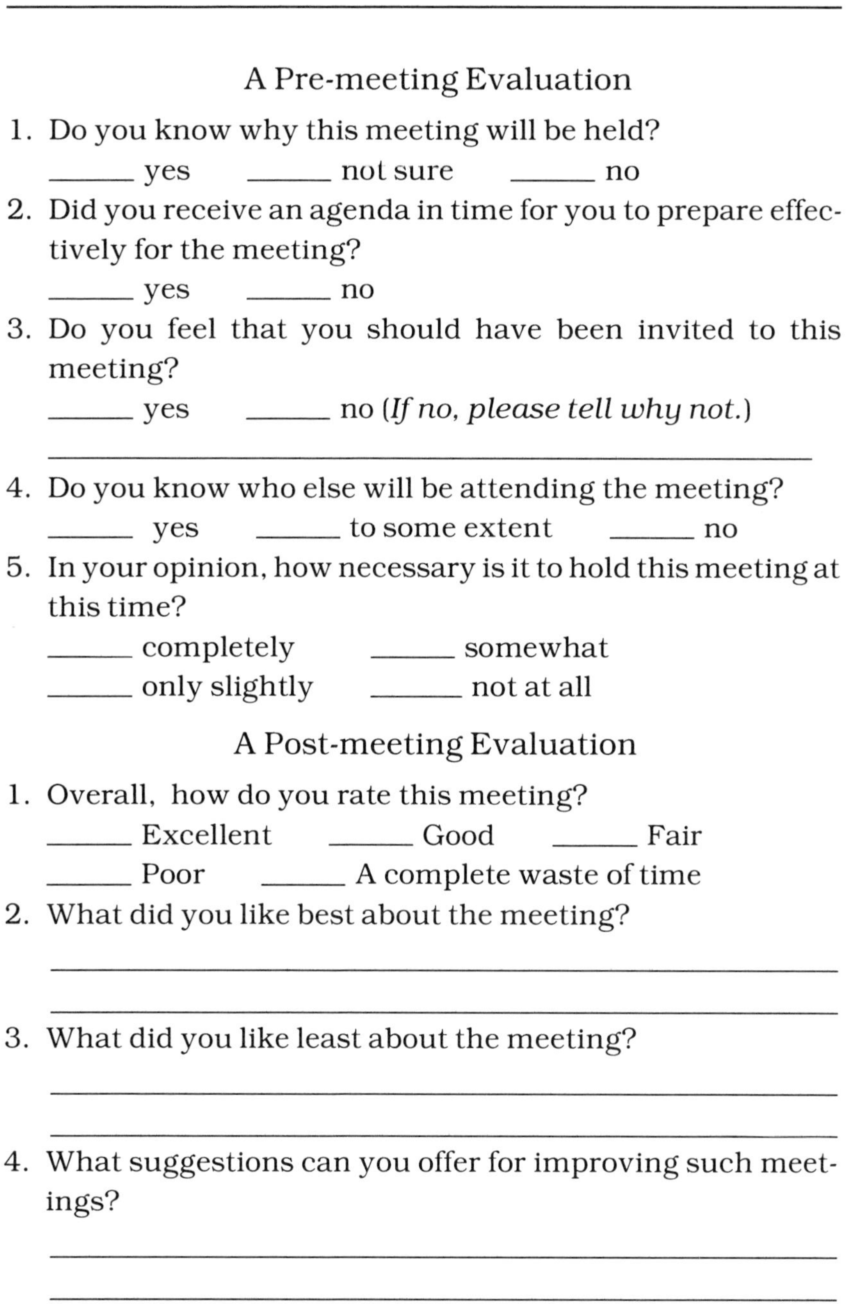

A Pre-meeting Evaluation

1. Do you know why this meeting will be held?

   ______ yes ______ not sure ______ no

2. Did you receive an agenda in time for you to prepare effectively for the meeting?

   ______ yes ______ no

3. Do you feel that you should have been invited to this meeting?

   ______ yes ______ no (*If no, please tell why not.*)

   ____

4. Do you know who else will be attending the meeting?

   ______ yes ______ to some extent ______ no

5. In your opinion, how necessary is it to hold this meeting at this time?

   ______ completely ______ somewhat
   ______ only slightly ______ not at all

A Post-meeting Evaluation

1. Overall, how do you rate this meeting?

   ______ Excellent ______ Good ______ Fair
   ______ Poor ______ A complete waste of time

2. What did you like best about the meeting?

   ____

   ____

3. What did you like least about the meeting?

   ____

   ____

4. What suggestions can you offer for improving such meetings?

   ____

   ____

4

Participating in Meetings

ONE of the more amusing recent articles on meeting participation lists several misbehaviors: the Heckler, the Know-All, the Whisperer, the Silent One, the Doodler, and the Hairsplitter.

Try this experiment: read that list again as you let your imagination take you to some previous time in your life. Where did your mental bus stop?

Mine parked squarely in front of third grade. There I am, in the fourth seat, third row. Bored. And even then my classmates and I (and perhaps you, too) were inventing BSTs (Boredom Survival Techniques) that we would practice through grade school, refine to an art in high school and college, then carry as basic mental habits into business meetings throughout our careers.

BECOMING AWARE OF YOUR OWN BSTS

You may be so good at surviving boredom that you're no longer conscious of what you do and how you look and sound in a meeting. That was the case for a group of Santa Fe managers in my 1987 summer communications workshop at USC. I had videotaped about 20 minutes of a long orientation meeting for another group, aerospace engineers unknown to the Santa Fe group. The camera was focused on the engineers—their facial expressions, movements, and especially their eyes.

I asked the Santa Fe managers to watch the tape, but with the volume turned off. Please jot down, I told them, the feelings you see being expressed by these people.

The resulting notes differed, but consensus was clear: "bored," "thinking of something else," "watching the clock," "unimpressed with the speaker," "sleepy," and "on a different planet."

Then I showed the Santa Fe managers the actual evaluation forms completed by the engineers following the orientation. These evaluations positively glowed with excitement: "excellent presentation," "clear and useful information," "good energy and enthusiasm on the part of the speaker."

Why the difference between what they were feeling on the inside and showing on the outside? They (and we) probably have fallen prey to old habits. Our BSTs, so well practiced for so many years, have become us—and, in meetings, they certainly don't become us. We've fallen back on these BSTs so often in the past that they take over almost without our will in any situation vaguely resembling a classroom. Put us in rows, at desks or tables, and let a speaker start to drone—that's all it takes. We switch on the autopilot of our own favorite boredom fighters.

What are yours? It may be none of my business, but you should make it yours. You are appraised by others, after all, primarily by what they see you doing, not what they hear you saying. If a video camera focused on you throughout the next meeting you attend, what would it reveal?

I'll assume that you aren't about to perform this telling experiment. The next best thing, then, is to observe someone else during your next meeting. What unattractive, disadvantageous signals is he or she sending by his or her own display of BSTs?

Take Cal across the table from you. The meeting is no more than 10 minutes along and already he's playing his BSTs like a string of worry beads. First he colors in all the O's on the agenda. Then he doodles something akin to a bat or spider web on the corner of the page. Tired of art, he starts the finger ballet—one nail cleaning or picking at another, fingers massaging knuckles, thumbs slowly circling each other. Jewelry comes next—the wedding ring has to be turned four or five complete revolutions. And the watch—Cal has probably had his watch since third grade. It must be taken entirely off the wrist, stared at for half a minute or so, then "wound"—though it's battery powered!

We'll spare you Cal's nose, ear, neck, and jowl play. But look for it at your next meeting. These signals paint a picture of Cal for each member of the group, including the meeting leader. Cal would be shocked by that picture—"I was paying attention! I was interested!" Perhaps so, but appearances told a different tale.

To assess your own BSTs, use the following checklist. Once you've recognized your own disadvantageous behaviors, make a concerted effort to fight the feeling. Undoing the habits of several decades takes resolution. The payoff, however, is substantial: a virtual make-over of your image in the eyes of others as a lively, concerned, focused participant—the kind that gets promoted.

Winston Fletcher has catalogued an immortal list of bizarre BSTs for *Savvy*. Here are three of my favorites:

- **Mark My Words** The object is to spot hidden agendas and interpret passive-aggressive jabs. For example, Alice's remark—'Why don't we consider alternatives?'—would translate into 'Anything but David-the-Donkey's plan.'
- **Slip 'n' Slop** Introduce a slip of the tongue to render participants weak and lower the general tone of the discussion. Television announcers often have fun this way, as in 'Rumor the president would veto the bill has come from a high White Horse souse.'
- **Invent the Proverb** Legend credits this game to Jeremy Bullmore, chairman of J. Walter Thompson's London division. Simply invent a phrase that sounds

Boredom Survival Techniques

(In nonverbal ways, we each "do things" at meetings that characterize us, often to our disadvantage, for other meeting members. The following list may help you locate and control your own Boredom Survival Techniques.)

Paper Play
Studiously coloring in the Os on the agenda
Shuffling papers or aimlessly paging through materials
Doodling
Reading (or writing) materials not germane to the meeting

Hand Play
Cleaning/picking at fingernails
Silently drumming fingers
Cracking knuckles or wringing hands
Twisting a ring or bracelet
Scratching and rubbing

Clothing Play
Picking at lint
Adjusting and readjusting collar, coat, sleeves, etc.
Pulling at buttons
Rubbing soft cloth

Eye Play
Staring blankly into space
Studying objects in the room
Closing eyes, as if concentrating (or napping)
Going through "waking" behaviors (rubbing eyes, stretching, rolling eyes, squinting)

Posture Play
Slouching in chair
Continual readjustment of chair or sitting position
Rolling head on neck

Miscellaneous Sports
Clearing throat over and over
Capping and recapping pen
Examining or feeling the texture of the table top
Endlessly fiddling with coffee cup, stirrer, etc.

rich with philosophical subtlety but has no meaning whatsoever. Bullmore's two old standbys are 'Somebody has to bury the undertaker' and 'It may not be the man who saws the logs who needs the fire.' "

HOW TO GET READY FOR A MEETING

Let's be optimistic—you've gotten a grasp on your own BSTs and are now ready to be "proactive" (a beloved business word) in your preparation for an upcoming meeting. Take it step by step.

1. Find out the purpose of the meeting. Easy, by a simple glance at the agenda? No. Agendas usually reveal topics, not purposes. Meeting topics are to meeting purposes as bales of hay are to a slaughterhouse. Find out what the meeting leader wants to accomplish in the meeting. Your boss, for example, may want to find out who opposes the postponement of the Brookview project. All the agenda says is "Discussion of Brookview Project"—but you know better.

One obvious way to discover the purpose of a meeting is to ask the meeting leader: "Barbara, what do you think will come out of the meeting tomorrow?" If you have a boss who likes to manage by mystery, you may have to be more subtle: "We're trying to get ready for the meeting tomorrow. Is there anything in particular you'd like us to think about?" Give the boss (or the boss's secretary) an opportunity to hint at the secret agenda of the meeting. It's not paranoid to believe that most meetings have them.

2. Find out who will attend the meeting. Often a distribution list on the memo announcement will note meeting participants. Or you may have to tap the secretary: "I'm preparing some handouts. Who's going to be attending that meeting tomorrow?" Don't be surprised if some major decision-makers aren't included on the distribution list but attend the meeting nonetheless.

Knowing the players is crucial to playing the game. Consider not only why these people were invited but also why others were left out. What does it tell you that the boss didn't invite anyone from the Budgeting Office?

3. Think about the agenda. I asked twenty managers to describe what they did with an agenda received a day or more before a meeting. The most frequently used word in their responses was "glance"—

- "I *glance* at it to make sure I'm not scheduled for a presentation."
- "I give it a *glance* before putting it in the file I'll carry to the meeting."
- "I usually *glance* at the time and place to make sure it fits my calendar."
- "I look at it just long enough to figure out what it is, then don't give it another *glance* until the meeting begins."

A *glance*? These managers have forgotten that their performance in meetings is a major determinant of their future with the company. Merely glancing at an agenda prior to a meeting is like buying a lot in Florida from a photograph. Take time to study the terrain.

Plan out what you want to say for particular agenda items. Decide what information you should have at your fingertips or what resource material you should bring along. Take care, however, not to put others off by your preparation. Stanley Bing writes in *Esquire* of the pitfalls of overkill. "Not long ago, a friend of mine relates, a newly appointed vice president at a brokerage firm was asked to make a brief presentation, a get-to-know-you chat, at a monthly meeting of the general staff. The new man came prepared to a fault. His spiel took thirty minutes. It began with some extensive personal background, then launched into his plans for his department, complete with slide show. 'By the end of that meeting,' recalls my friend, also a vice president, 'the guy was no longer viable. We just wanted to say hi, and he went into this orgy of self-promotion. Who wanted to hear all that stuff?' "

4. Think about the time and location. You can probably guess much about the mood and pace of the meeting by considering when it will be held. A 9 A.M. meeting? Prepare for a brisk pace, fast-moving discussion, and quick votes. Just after lunch? Bank on a more relaxed pace, a willingness of the group to endure wandering discussion, and a pervasive gloom over prospects and projections. A Friday afternoon meeting? It's one of Goldie Brass's favorite times to call an action meeting at Shoppers Drug Mart: "People will attend to business and stick to the point because it is Friday and they want to go."

Location can also help prepare you and help you prepare for the meeting. If the walnut-panelled boardroom has been chosen, dress up. For presentations there, choose your most professional visual aids, even if it means leaving behind your favorite flip chart. A more casual location will obviously dictate other choices: less formal dress and less impressive (but perhaps more expressive) visual aids. Some meeting participants make it a habit to visit the meeting site prior to the gathering, just as golfers like to walk the course prior to a tournament. In both cases, familiarity and planning can pay off.

HOW TO PARTICIPATE IN A MEETING

First, let's get you seated advantageously. Psychologist John Dovidio, a specialist in nonverbal communication, points out that dominant figures at meetings (leaders and other key players) typically select the ends of a rectangular table. When, by a random shuffle, more withdrawn people find themselves sitting in those positions, they often speak up much more than usual.

But let's say you're not one of the kingpins or queenpins. By sitting right next to the meeting leader, you minimize your effectiveness as meeting participant. You can't easily look the meeting leader in the eye as you speak (you're too close) and the leader may often look past you to call on others. Find a seat, therefore, where you can comfortably address both the chair and the rest of the meeting participants. Don't worry if you're sitting right next to a key opponent of your ideas. Interpersonal friction tends to be reduced by proximity—in other words, if you're close enough to hit your enemy, you'll probably be on good behavior and so will he or she.

The meeting begins and so far you haven't said a word. Not smart. Research in group dynamics demonstrates that early speakers will tend to contribute more often and more fully to discussion than those who do not speak up early. The explanation for this phenomenon is partly physiological. The excitement (or nervousness) of speaking up sends a burst of adrenaline through the system—most of us have felt this sudden "wake up" when we begin to speak before a group. That stimulation makes you a keener listener and faster thinker for the ensuing moments of the meeting. By speaking early, you've physically given yourself a leg up on other meeting participants.

You've also bonded early as part of the "in" group in the meeting. Those who don't take the risk of speaking up, who remain as spectators on the sidelines throughout the meeting, have to overcome the stigma of "outsider" when they do finally offer opinions. Notice in this regard that effective meeting leaders usually hold the floor for a minute or more at the beginning of the meeting, often for the ostensible purpose of giving "background" or "opening remarks." Psychologically, they are using this early occasion to assert themselves as primary figures in the emerging verbal action. In effect, they want to be first in the door of the clubhouse. Follow them!

Consider putting a bit of spin on your content when you speak up in a meeting. Richard Wiegand, professor of management communications at the University of Alabama Graduate School of Business, warns meeting speakers against the belief "that they have to fill the audience in on all the background before the major idea . . . can be appreciated. . . . Unfortunately, the group is often half asleep before the big point is revealed."

Wiegand recommends that meeting participants cut to the chase and get to the point right away. "It's a real time-saver if the speaker starts off with the conclusion. Most presenters find that once they have voiced the key idea, fewer details are needed." For example, if you want to talk about the results of a recent marketing drive you might begin with "We found two marketing channels that work, and one that doesn't." The group will certainly prefer the clear, concise details that follow from this statement to the usual alternative: "As you know, my group has been active in the development of blah, blah, blah . . . leading to an investigation of blah, blah, blah. . . ."

But all is not sweetness and light. You know that the comments you're about to make will upset another member of the group. Walter Kiechell III recommends in *Fortune* that "if you suspect that your assertions on a particular subject may gore another participant's ox, or even nick it, tell him in advance that you are going to raise the issue." This notice lets the participant know that you're not trying to launch a surprise attack. When you can't make this contact prior to the meeting, use verbal "bridges" to show how you intend your viewpoint to be understood in relation to the other participant's position:

> "Jack's point about EEOC requirements is well taken, but we can achieve those same goals by . . ."
>
> "We're trying to see this issue from as many points of view as possible. In addition to what Jack has offered, we should look at . . ."

In such moments, especially when a subordinate speaks against the position of a boss, it's necessary to choose words that allow everyone to save face. You can, for example, explicitly state that you're "playing devil's advocate"—a role most business leaders have come to appreciate. Or you can state your view as a "related perspective" or "follow-up idea," even when you—and the group—know you're meeting another position head-on.

"Could we blame the CIA?"

For more extended contributions, take the time to develop appropriate visual aids. (Bear in mind, though, the dangers of overkill here.) B.Y. Augur, vice president of the Visual Productions Division at 3M, likes to point out that "when relying on verbalization alone to communicate, an estimated 90 percent of a message is misinterpreted or forgotten entirely. We retain only 10 percent of what we hear. But adding appropriate visual aids increases retention to approximately 50 percent."

The most quoted rule of thumb for such visuals also comes from 3M: *only one essential concept per visual; no more than six words per line; and no more than six lines per visual.* Here's what a slide or transparency looks like when following these approximate guidelines:

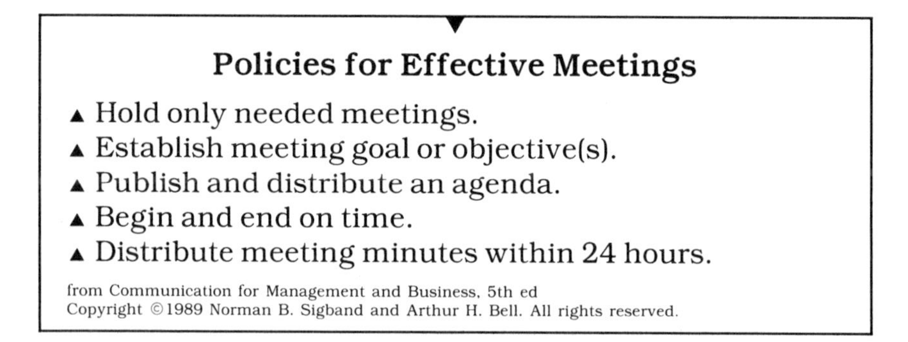

So far we have you well seated and well spoken. But most of the time you're listening, not speaking—and, like most of us, you've probably never had even an hour of training in this crucial business skill. (How ironic, in fact, that schools do not provide instruction in the very thing that students are supposed to do for twelve or more years!)

As others speak, show that you are interested in their points. Look at them, for example—a radical idea for some meeting participants. Demonstrate by an occasional nod or frown or smile that you are thinking along with them, that you follow their train of thought. Such attention doesn't signal that you necessarily agree with what's being said, but simply that you're taking it seriously.

One immediate result of such active listening is the attention that will flow toward you from the speaker. Notice that when you speak you tend to come back to friendly, interested faces again and again. The traditional name for this bond of interest is "rapport," and it can hardly be underrated as a mover and pusher in business relations. Listening actively and attentively even to your business opponents goes a long way toward improving mutual working relations. Rapport, in other words, breeds rapprochement.

Taking brief notes may help you and compliment the speaker. "But," says James D. Kimes, vice president at Vestal Laboratories, Inc., "don't write a book. You can't record everything that is said and done at a meeting, but you can write down salient facts, even if it's just a word to remind you of a point. At some meetings I have observed people writing with such sustained zeal that one would suspect they were writing a letter or a memo that had nothing to do with the meeting. It is difficult to listen with understanding and take notes, too, so if a comprehensive record is necessary, a secretary should be invited to take notes so everyone else can concentrate on listening."

Finally, ban tape recorders and VCR cameras from your meetings. At the click of the recorder, otherwise thoughtful partipants freeze in their verbal tracks or, just as bad, turn into Perry Mason. (Have you noticed what TV cameras have undone for the U.S. Senate and House?) Besides, as Kimes points out, "recordings are too permanent for some. You can always dispute what someone thought you said, but having it down on tape scares people."

HOW TO FOLLOW UP AFTER A MEETING

By a stroke of good fortune, the meeting adjourns on time. But the meeting isn't over. Its business and central concerns will be carried through by those who have chosen (or have been assigned) tasks valuable to the work of the group. That's you. It was easy to see, after all, that those who took on follow-up tasks got to associate more directly and more often with company leaders.

Doing a good job on such tasks gets you noticed by people who can help you. Dartmouth business professor Leonard Greenhalgh points out that you've put yourself in a favorable position for future meetings: "you can report back to the people at the meeting with your information and also give yourself a good claim to the outcome."

It's also a good idea to give feedback after the meeting to other participants. You liked Jill's presentation? Tell her so. Frank's comments made you reconsider your position? Let him know that he gave you food for thought. Bob's handouts were superb? Ask him if you can make copies to pass out to your people. In Zen philosophy, this is the "sound of the other hand clapping." In more common terms, it's simply the other shoe dropping. Let participants know that their words didn't disappear into a vacuum, but instead made a difference for you. No one expects you to play Florence Nightingale in giving T.L.C. to your fellows. The point is simply that team members need to talk to one another in business.

To assess your own development as a meeting participant, use the following checklist after each of the next four meetings you attend. Locate areas of strength and areas for improvement, then make specific plans for your own professional growth as a meeting participant.

Measuring Your Own Participation

Meeting participation is a matter of quality more than quantity. Hence you can't evaluate your meeting skills by simply counting how often you spoke up. But you can gain valuable insight by determining your participation tendencies, guided by the following list. It is best to fill out this self-evaluation immediately following a meeting. For interesting comparison, ask a colleague also to fill out this estimate of your meeting behavior.

1. Where I sat (*sketch*):

2. How this seating position affected my participation:

   ____

   ____

3. Types of participation: (*Place numbers in each blank according to the following scale: 1 Agree 2 Not sure 3 Disagree.*)
    ____ I spoke up to initiate new ideas.
    ____ I spoke up to disagree with other speakers.
    ____ I spoke up to agree with other speakers.
    ____ I spoke up to offer additional information.
    ____ I spoke up with questions.
    ____ I spoke up to clarify points for others.
    ____ I spoke up to summarize points for others.
    ____ I spoke up in a humorous way related to the point at hand.
    ____ I spoke up in a humorous way unrelated to the point at hand.
    ____ I did not speak up.
    ____ Other: ______________________________

4. My attitudes towards my participation. (*Place numbers 1, 2, or 3 as in exercise above.*)
    ____ I felt I expressed myself clearly and persuasively.
    ____ I felt I expressed myself clearly but not persuasively.
    ____ I felt that I did not express myself clearly or persuasively.
    ____ I felt that the group usually let me have my say.
    ____ I felt that I was usually cut off by other participants before having my say.
    ____ I felt that I spoke up too often.
    ____ I felt that I wanted to speak up more often, but did not.
    ____ I felt that I spoke up too little.
    ____ I felt that I communicated an impatient, somewhat angry attitude to the group.
    ____ I felt that I communicated a discouraged, cynical attitude to the group.
    ____ I felt that I communicated an upbeat, energetic attitude to the group.

What to do with your results: This self-evaluation does not yield a "score" so much as a self-portrait of your meeting behaviors. Review your responses to decide what areas, if any, you wish to change. Compare your self-evaluation with an outside evaluation by a colleague. Discuss those areas where your self-perceptions differ from the perceptions of others.

Julie Bailey, a former editor at *Business International*, recently published the following suggestions on "How to Stand Out at Someone Else's Meeting" in *Working Woman*:

> You're not running the show, but that doesn't mean you can't stand out. Here are a few tips to make you a sought-after participant:
>
> ▲ **Do your homework—before the fact.** When an agenda hits your desk a few days before a meeting, read it. Make notes about relevant points to discuss next to each entry. If you have questions, ask them sooner rather than later. The more information you have compiled before entering the boardroom, the better.
>
> ▲ **There are no small parts, only small actors.** This holds true in the workplace as much as it does onstage. There's no need to hog the limelight. "Don't be tempted to grandstand," cautions marketing consultant Marlene Rossman, adding that there's no need to threaten the chairperson's authority by attempting to gain control. Instead, hold your thoughts until the time is perfect. Then let them wow your co-workers.
>
> ▲ **One great idea is better than a handful of so-sos.** Limit idea-making to one or two issues where you can back up your proposals with statistics, experience, or new research, Rossman suggests.
>
> Try to test the reaction to these issues before the meeting if you can, says Lucille Corrier, director of communications and marketing for Touche Ross Financial Services Center. Talk to your co-workers, even your superiors. "With a new idea I'm

very up front about going to my boss and asking if he will support me in the meeting," Corrier says, adding that she would rather rework her proposal beforehand than risk the prospect of a no-go vote in front of her peers.

- **When you're the new kid on the block, it's important to follow someone else's lead.** Stand back and let people assume their regular seats before you take one. You don't want to lose the support of the senior vice president by taking her favorite chair. Observe a meeting's dynamics before jumping into the action. Your old company may have encouraged screaming matches, but these people could be whisperers.
- **All ears? Three cheers.** Listening attentively—with your eyes and your body as well as your ears—can win points, not to mention the same consideration when you've got the floor. If you're whispering to your neighbor, you'll never know when important information is flying past. Besides, that's grade-school stuff.
- **A cause-and-effect relationship.** "Link your ideas to a promise of action and the prospect of immediate benefits," suggests Communispond VP Paula Ludlow. Show that A leads to B, which leads to C, using charts, diagrams, and past examples.
- **"No one can resist a volunteer,"** says Ludlow, citing her experience at a meeting to discuss marketing the firm's business-communication courses. The majority of her coworkers thought course prices should be raised. But Ludlow suggested lowering them by 15 percent for groups of 5 or more students. To counter skepticism, Ludlow offered to have her department carry out a limited direct-mail campaign. When the strategy worked, her idea was adopted.
- **Don't drop the ball.** If you've been asked by the chair to investigate an idea or develop a proposal, do so. Send a memo summarizing your findings within several days. Even if you haven't been able to finish the job, send a memo to let her know that you're working on it.

ROLODEX RECAP

▲ discover the Boredom Survival Techniques that may be communicating mixed or wrong messages about you to other meeting participants

▲ prepare for a meeting by finding out its purpose, attendees, agenda, time, and location

ROLODEX RECAP

▲ participate in a meeting by sitting in an advantageous place, making your points straight-forwardly, listening actively, and using visual aids where appropriate

▲ follow up by giving feedback to participants, including the chair; taking on tasks assigned in the meeting; and assessing your own participation behaviors

5

Small Meetings

"WHEN it comes to meetings, companies should often think small." That advice, says Georgetown University management professor Dayle Smith, flies counter to the "bigger is better" philosophies of many American corporations.

"But in meetings," says Smith, "companies have much to gain by thinking small. Meetings of two, three, or four employees instead or ten or twenty tend to be more focused, more action-oriented, and less time-consuming. For all these reasons, small meetings save the company money."

How small is small? Consider the smallest possible meeting, short of talking to oneself: the one-on-one meeting. Taking those common encounters seriously as small meetings brought a new level of business efficiency at Intel Corporation. President Andrew S. Grove finds it curious that "when I ask a manager from another company [about one-on-one meetings], I usually get an 'Oh, I don't need scheduled meetings with my supervisor [or subordinate]. I see him several times a day.' "

But at Intel, one-on-one encounters aren't treated casually. They are defined, in Grove's words, as "the principal way your business relationship is maintained. By talking about specific problems and situations, the supervisor teaches the subordinate his skills and the Intel approach to things. At the same time the subordinate provides the supervisor with detailed information about what he's doing and what he's worried about."

ONE-ON-ONE MEETINGS

Like larger meetings, the one-on-one meeting should be scheduled in advance on both parties' calendars. "It may sound like overkill," says Seattle heavy equipment broker Ray Thompson, "but it really helps to know even a day or two in advance exactly when I'm supposed to meet with someone and for how long. It gives me a chance to prepare." Thompson says that scheduled meetings have less chance of "cloning" than unscheduled meetings: "We used to get together on an unscheduled basis only to find out what we needed to bring to the next meeting. With scheduled meetings, that second get-together is usually unnecessary."

The time allotted in advance for one-on-one meetings depends, of course, on the work to be accomplished. In general, however, business men and women discover that the 30-minute time period works well as a target. Both parties are "under the gun" to get to the point. At many companies using the 30-minute slot, employees are urged to schedule in an additional 15-minute margin in case meetings need to run longer.

Matching your meeting time to the personalities involved can be more a matter of art than managerial science. "Everyone operates on a different metabolic clock," says Sonya Hamlin in *How to Talk So People Listen.* "To get the best from people, try to take their energy patterns into account as you schedule your meeting. Find out enough about the other person's work habits from a secretary or colleagues to know what time of day to call the meeting. Give yourself a break by selecting your best time, too."

Also specify the place for one-on-one meetings. Each of us defines our own office space as a den of sorts—our "safe place" where people don't rearrange furniture, open drawers, or borrow desk items without our permission. We mark this secure space with our name—on the door, the cubicle wall, or the desk. When choosing to meet one-on-one with someone, we know we will gain these advantages by meeting in our space instead of theirs:

- ▲ we will feel more comfortable, "at home."
- ▲ we will tend to be more assertive in discussion, more empowered.
- ▲ we will have more information at our fingertips. After all, our files, phone, FAX, and computer are all within reach.

At the same time, we can expect the other person to feel less comfortable, less dominant than usual, and somewhat disadvantaged with regard to information resources.

The implication is clear: steer the meeting toward your own office when you want to drive the bus. When meeting in someone else's space, recognize your territorial disadvantage and compensate for it by (a) psyching yourself up for an energetic defense of your points and, (b) taking along any support materials you may need.

There are many times in business life, of course, when you'll want to put a bit of reverse spin on the territorial realities, giving the other person more confidence, more respect, more access to resources during discussion. As a boss, for example, you may want to build a subordinate's sense of independence and self-initiative by meeting in his or her office. Or, by meeting there, you may simply want to shift the burden of meeting leadership (in terms of preparation and organization) to another person.

Intel's president, Andrew Grove, says a manager "can learn a lot simply by going to his subordinate's office. Is he organized? Does he repeatedly have to spend time looking for a document? Does he get interrupted all the time? Never?" Or a manager, especially one who practices MBWA—Management by Walking Around—may want to meet in employees' offices as a valuable way to build team spirit and a sense of distributed decision-making.

When the expense account can bear the surcharge, consider off-site locations for particularly important one-on-one meetings. "When we get away from the office for a meeting, we're often different people in terms of attitudes and creativity," says Devon Scheef, TRW's Director of Training. Off-site locations can include restaurants, conference rooms, and clubs. Think twice (at least) about meetings at nearby watering holes—the atmosphere and alcohol may subvert business goals.

That's not to say that productive meetings can't blend fun with business. Stanley Bing writes of a friend who "accepted an invite to spend a week with his chairman and six key managers at a tiny, plush resort in Vail. They hit the slopes at dawn, ate and drank to excess, and sorted out operational priorities around a toasty fireplace at night. Next year, he'll go again."

One perennially popular time and place for one-on-one meetings is at breakfast. These "power meals" take place each workday morning at such prestigious locations as New York's "21" Club and the Rockefeller Center Club, where membership and dues can run into the thousands each year for a breakfast meeting spot. At less airy heights, the tables of Denny's and Big Boy are filled from 7:00 to 9:00 A.M. in large part by business tête-à-têtes. But whether upscale granola or corn flakes, breakfast does not always a meeting-of-the-minds make. Other diners can sour your agenda. For Donald Beldock, chairman of Basix Corp., who has breakfasted at the top, "the difficulty with those kinds of places is that one tends to be distracted by seeing a great many familiar faces and getting up to shake hands. . . . It does tend to dilute the amount of business that can be accomplished."

It's a good idea in advance of any one-on-one meeting to settle on the question of ownership—whose meeting is this, anyway? Take the case of a supervisor meeting individually with six subordinates. If each one-on-one meeting is defined as the supervisor's meeting, the supervisor will end up preparing for six separate meetings. If, on the other hand, the supervisor shifts the burden of meeting ownership to the subordinates, his or her need to prepare is lessened considerably.

The Meeting Outline

Whoever owns the meeting has the obligation (the "monkey," as some companies call it) of preparing an outline of items to be discussed. If you're not used to formalizing one-on-one meetings, this obligation may seem nit-picky. An outline for a simple half-an-hour meeting with Linda?

Yes, if it's Linda's meeting and you want to participate effectively. You need to know the items on Linda's agenda, and the order in which she wants to discuss them. You have a chance to watch in advance for what Intel's Andrew Grove calls "zingers"—hot items, often of a heart-to-heart nature, that usually get stuck at the end of an agenda even though they deserve top priority. For example, Grove says, "the subordinate may suddenly announce he's unhappy and has been looking outside for a job, giving you only five minutes to deal with the issue. You should scan the outline at the beginning of the meeting, look for these, and rearrange the order of subjects to bring them forward."

An outline for a one-on-one meeting is usually less formal, but no less thorough, than an agenda for a larger meeting. Notice in the following example, communicated as a memo, that items are logically arranged, clearly described, and include some indication of support materials needed for the meeting. (The memo, by the way, took Sylvia Lopez six minutes to write—certainly a good investment toward the efficiency and effectiveness of the meeting.)

To: Vincent Forge
Supervisor, Personnel Accounting
From: Sylvia Lopez
Vice President, Operations

As we discussed, Vince, let's meet from 9:30 to 10:00 or so in your office on Thursday, May 7, to discuss the following:

▲ recent upswing in personal absence days. (Please bring along any hard data you have—I want to understand the causes.)
▲ preparation for the July 31 audit. Do your people need anything from me?
▲ some crystal-ball forecasting of hiring needs in your department for next year. I know you don't have a firm fix

on this question yet, but the more I know the more I'll be in a position to help you when the time comes.

Give me a ring (#2833) if you have questions or want to include other items.

The tone of this memo suggests—and may help to determine—the tone of the meeting itself: cooperative, open to new ideas, but at the same time focussed and organized.

STAFF MEETINGS

"Many of the weekly staff meetings that I attended," says Dennis Griffin, Senior Systems Programmer at First Wisconsin National Bank, "weren't unnecessary—they were just unnecessary for *me* to attend."

There's an inevitable tendency to invite too many people to a staff meeting. The reasons are understandable. First, we've prepared hard for the meeting and wouldn't mind at all if we had a wide audience. Second, the meeting is a sign of activity on our part—and it would be good if a lot of people saw us in action. Finally, there's no end to the question of "need-to-know"—everyone in the company could potentially be involved, at least marginally, in some of the topics discussed at the meeting. So why not invite a crowd?

The question of expense aside, inviting too many of the wrong people to a meeting is a formula for disaster. The ability of a group to reach a decision—or to do any serious work at all—is in direct proportion to its size and homogeneity. Jane Lockshin, president of Custom Databanks, Inc., is careful to select participants who want and need to work together. "Too many different political agendas," she says, "change the original purpose of the meeting."

Assembling a productive group often involves an evaluation of the egos and levels of corporate power in the room. Given the spoken and unspoken rules of company civility and servility, you can expect rigor mortis in meetings with the wrong mix of big chiefs, little chiefs, would-be chiefs, and plain old warriors. Geraldine Henze, director of communications at Columbia Business School, advises "never have

"I see we're having our monthly sales review."

Reprinted by permission of *Sales & Marketing Management*. Copyright 1986.

more than two hierarchies of management present" at the same meeting. "If you do, some participants are tempted to grandstand. Others just clam up."

The Meeting Agenda

The agenda for a staff meeting should be circulated at least five business days prior to the meeting. It's often a good idea to get input from key decision-makers before distributing a trial agenda. Eleanor Reynolds, a partner at Ward Howell International, makes it a habit to pass a tentative draft of her agenda to key players before finalizing it for a meeting. "I know the reception around the conference table will be warmer," she says.

The ideal agenda specifies time, place, attendees, contact person, purpose, items for discussion, time allotments, and key input people. That rather long shopping list seems less intimidating when viewed as a single-page example:

Agenda

Staff Meeting: 9 A.M. – 10:30 A.M., June 8, 1989
Place: Conference Room A

Attending:
Ruth Evans, Marketing
Bill Philips, Accounting
Tom Wilkins, Accounting
Toyo Su, Advertising
Jean Peters, Advertising
Art Brown, Operation

Information contact: Lisa Shore, New Products, ext. 2141

Purpose of the meeting: to update participants on three new product lines slated for 1990–91 cycle and to discuss any obstacles or problems these pose to your individual areas of responsibility

Suggested preparation: review proprietary new products files on reserve at company library desk

Item	Suggested Time	Input Person	Action Taken	Action Person
1. Update on FDA approval of Swanzo detergent	10 min.	Bill P.	*(this portion filled in during meeting)*	
2. Expected delays on lotion ads	5 min.	Jean P.		
3. Discussion of mail marketing for designer creams	20 min.	Toyo S.		
4. Budget impact of postal rate increases	20 min.	Tom W.		
5. Prelim. market data for color-safe bleach	30 min.	Ruth E.		

This agenda, arriving on each attendee's desk, announces loud and clear that the meeting won't be a free-for-all or a waste of time. Participants know who to contact if they have questions before the meeting. Key individuals who will be expected to speak up with reports and other input are given notice in time to prepare. They also know how long they should speak, judging by the total amount of time suggested (but not insisted upon) by the chair. There's room on each person's agenda to note what the group decided or discussed with regard to each item. (The minutes, received a day or two after the meeting, can be compared with these notes for accuracy.)

Donald Pine, Manager of Sustaining Engineering at Bull HN Information Systems, tells of an infamous meeting he attended in which "an hour was spent clarifying the purpose of the meeting and determining that certain individuals who were needed were not there and that some people who were there were not needed." An agenda can and should function as the meeting planner's "guide to the players." Each agenda item should be reviewed with an eye toward which employees should attend.

The Meeting Announcement

For especially important meetings or those not called on a regular basis, you may want to attach a meeting announcement to the agenda. This notice, usually in the form of a memo, tries to explain the purpose of the meeting in a way that motivates those invited. The tone of the memo should mirror the tone you want for the meeting itself.

To: (distribution list here)
From: Lisa Shore, New Products
Subject: meeting on three new products, June 8
Date: June 1, 1989

Thanks for making room in your schedule for this important meeting. You've been selected by your supervisor as a key person within your work unit to shepherd three new product lines to market in 1990–91. At the meeting, we'll share as much information as we have now about these products and discuss how each of our areas is impacted by them.

I think you'll especially enjoy Toyo Su's preliminary video on the ad campaign for our designer creams. It's definitely a new departure for the company!

Call me if you have questions about the agenda or suggestions for the meeting. See you at 9:00 a.m. on June 8 in Conference Room A.

The location for your meeting, in this case, is a company conference room. But for some occasions, you may want to stretch the budget a bit for the purpose of stretching the imaginations of participants. Marna Petersen is manager of meetings and conferences for Meeting Planners International. By finding interesting off-site locations for your small meetings, she says, "you're putting people in a more relaxed, casual setting. You've changed the pace for them. People are able to focus on the matter at hand without the distractions of the normal office routine. In this environment, they will be able to think more creatively and approach problems from a new angle."

As a general rule, says Loral Defense System's Ralph Franke, "the more plush the surroundings, the more seriously the subject matter of the agenda is taken."

The possibilities for off-site meetings are almost limitless. But the following quick walk-about of five regions may inspire you to investigate likely meeting spots near your company.

Barefoot in the Park Many national and regional parks have meeting facilities, including overnight lodging, that can be had for very little expense. David E. Gackenbach, chief of concessions management, U.S. Department of the Interior, says that "the National Park Service opens the doors to the parks' lodging facilities for conventions and group meetings as long as such usage is not environmentally damaging to the parks' resources and does not preclude use by the general public visiting the park." Meeting rooms at Yellowstone, for example, are available without charge if the group also takes overnight lodging.

Big Wheel Keep on Turnin' Steamboats such as the Delta Queen and the Mississippi Queen offer meeting space for as short as a day or as long as a week. The paddle wheelers move slowly along a route from St. Paul to New Orleans, on the Mississippi River, and from Cairo, Illinois, to Pittsburgh on the Ohio River. Group discounts from 15 to 30 percent are available, according to the time of the year. "Some of America's biggest companies got their start doing business aboard paddle-wheel steamboats in the 1800s," points out Edward B. Holbert, vice president of sales development for the Delta Queen. "Our steamboats carry on this tradition by providing a thoroughly modern meeting facility while traveling through the heart of America at eight miles per hour."

Riding the Rails Amtrak will rent you a car or an entire train for your meeting. The Air Force Association recently took that option in reserving several cars for a trip from Virginia to New York. Paul Nestor, mid-Atlantic district sales manager for Amtrak, sees the train as a superb meeting location. "It creates a unique mood. Whatever your needs are, we can sit down and work it out."

Imagine a Mansion Within driving distance of most corporate headquarters are private mansions that can be rented out for meetings. The former mansion of tobacco king Bowman Gray, Sr., recently opened as the Graylyn Conference Center near Winston-Salem. The 50-year-old mansion features an indoor swimming pool in Art Deco style, 15th-century sculpted stone doorways, and dozens of rooms for small and large meetings. After the Eye Bank Association of America met there, president Tom Moore gave the setting a high rating: "An environment like this is perfect for coming up with ideas and concentrating on what you're doing. It was a peaceful atmosphere to conduct long-range planning and to set goals."

The Muses in a Museum By joining the National Trust, companies can gain a 20 percent discount for rental of Trust properties, according to Beverly A. Reece, public affairs officer. These include such stunning meeting sites as the 69-year-old Woodrow Wilson House in Washington, D.C.; with a meeting room for 25 meeting-style seats or 45 theatre-style seats. The National Trust lists hundreds of such treasures throughout the country.

The best sources for detailed descriptions of unusual meeting sites, including rates, are the following:

America's Meeting Places, published by Facts on File, Inc., 460 Park Ave. South, New York, NY 10016 (212) 683-2244

Places, 1988–89 edition, specializing in New York-area meeting sites. Available from Places, Box 810, Gracie Station, New York, NY 10028 (212) 737-7536

The National Parks Visitors Facilities and Services, published by the Conference of National Park Concessioners, Mammoth Cave, KY 42259 (502) 773-2191

Country Inns and Back Roads, available through Berkshire Traveller Press, Stockbridge, MA 01262 (413) 298-3636

SEATING OPTIONS

Negotiators have been known to argue about seating arrangements for days at peace conferences (such as the notorious Vietnam Peace Talks) and management/labor disputes. These men and women knew how important seating could be to the relative power of participants, to their degree of participation, and hence to the outcome of the meeting.

No single seating arrangement is ideal for all purposes. Each pattern comes with built-in advantages and disadvantages. The following patterns are common options for meetings with 12 or less participants. Consider the probable effect of each in deciding which best fits your audience and your purpose.

Theatre Seating

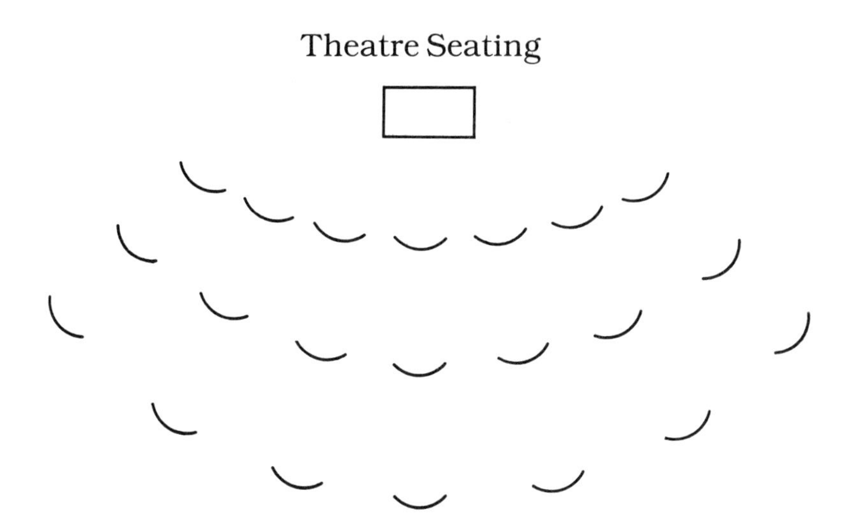

Effects on the meeting:

- ▲ emphasizes importance of leader
- ▲ provides audience with clear view of visual aids
- ▲ severely restricts participant-to-participant discussion
- ▲ sets a formal tone to the discussion
- ▲ makes note-taking difficult
- ▲ increases likelihood of leader-participant discussion through question and answer

Circular Seating without Tables

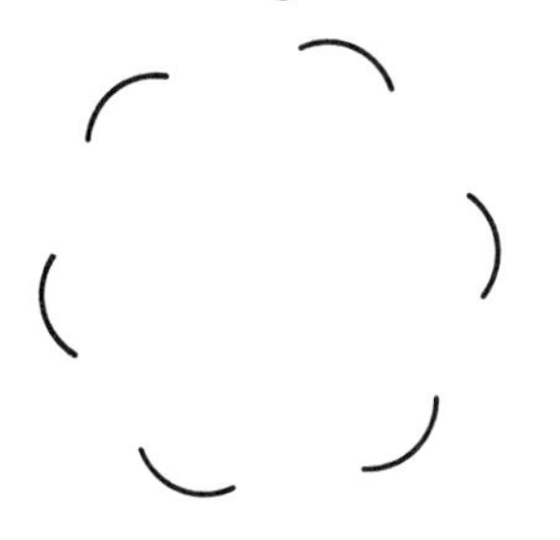

Effects on the meeting:

- ▲ de-emphasizes importance of leader
- ▲ reduces likelihood that participants will refer to notes or resource materials

- ▲ makes note-taking difficult
- ▲ encourages participant-to-participant discussion
- ▲ makes viewing of some visual aids awkward

Circular Seating with Tables

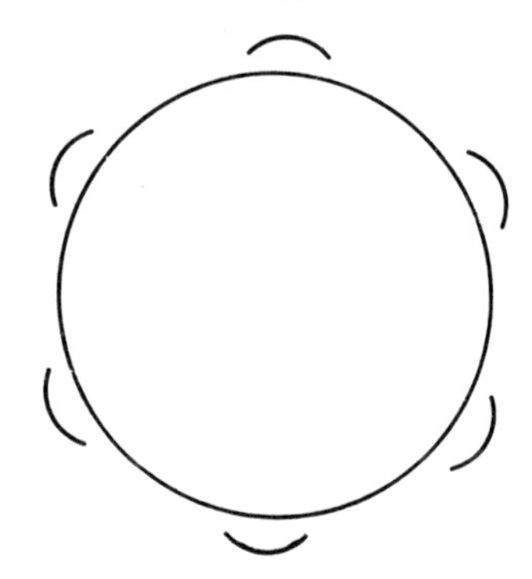

Effects on the meeting:

- ▲ de-emphasizes importance of leader
- ▲ makes viewing of visual aids awkward
- ▲ invites participant-to-participant discussion
- ▲ sets an informal tone to the discussion
- ▲ makes note-taking easy (as do all of the following table configurations)

Rectangular Seating

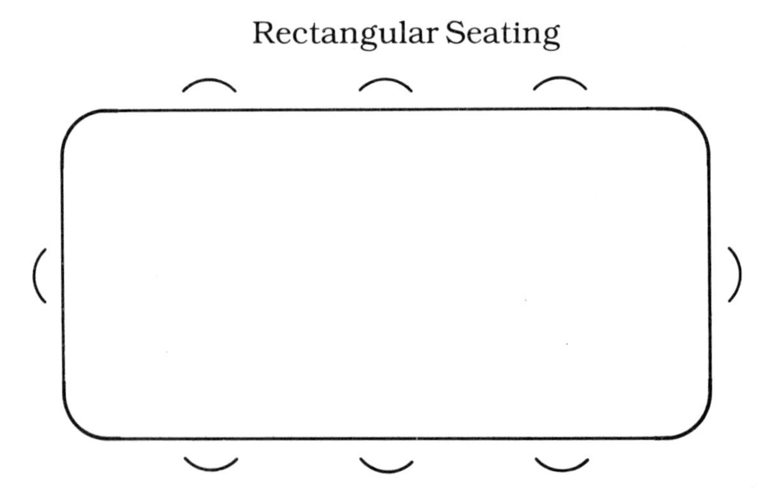

Effects on the meeting:

- ▲ emphasizes importance of leader and, at the other end of the table, the leader's chief cohort or antagonist
- ▲ invites formation of teams, with one side of the table opposed to the other side
- ▲ encourages some participant-to-participant discussion, especially across the table
- ▲ works well for viewing most visual aids

Oval Seating

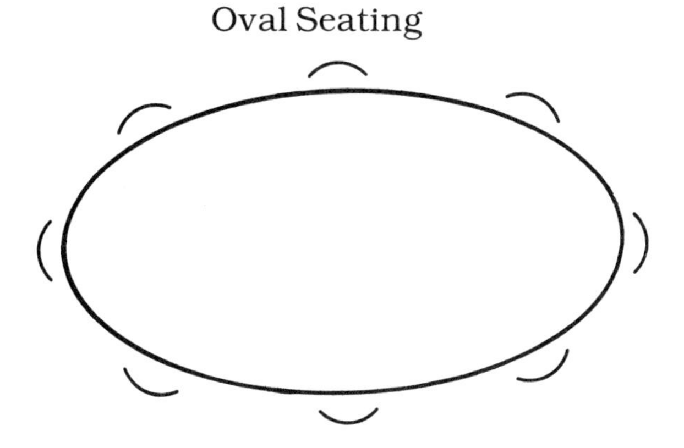

Effects on the meeting:

- ▲ all of the effects of Circular Seating, but with greater emphasis on the importance of the leader

Small Table Seating

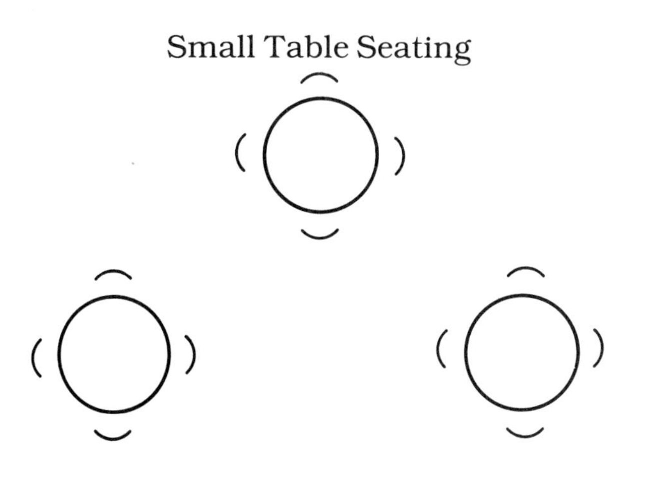

Effects on the meeting:

- ▲ encourages strong team identity and discussion for each table
- ▲ reduces importance of leader
- ▲ reduces likelihood of table-to-table discussion
- ▲ makes viewing of visual aids awkward
- ▲ sets an informal, restaurant-like tone for meeting

FOLLOWING UP ON THE MEETING

You've chosen your attendees with care, distributed a clear, concise agenda, set up the meeting room to support your purpose, and conducted the meeting according to the principles of Chapter 3 (or Chapter 4, if you are a participant). You're within striking distance of that rare experience, a good business meeting—but don't blow it now. The success of your meeting will depend not so much on its many strands of discussion as the knots you are able to tie at the end.

Sum Up the Accomplishments of the Meeting Use the last few minutes of the meeting to draw together what has been done and remains to be done. If possible (and it usually is), leave the group feeling that their efforts have been worthwhile and appreciated.

Review and Distribute the Minutes You've wisely delegated the taking of minutes to a secretary who can think, not just take shorthand. His or her notes intelligently and concisely reflect the discussion and decisions of the group, with key parties accurately identified. (You didn't make the mistake of asking a co-worker—one who might want to contribute to the meeting—to also take notes for the minutes.) Even when they are sent out above the secretary's "Respectfully Submitted" signature, the minutes for the meeting are your responsibility. If they are misstated, meeting participants can't help but wonder if the leader's influence didn't play an editing role after the meeting. Take time, therefore, to review minutes carefully and only then distribute them to meeting participants and other stakeholders in meeting issues.

Timing can be crucial, especially when the minutes serve to name individuals to particular tasks or to announce important decisions to the company at large. As a general rule, minutes should be distributed within two business days of the meeting. If the practice is maintained within your company, also post the minutes on a bulletin board (perhaps a computer bulletin board) for general review.

Give Feedback to Meeting Participants. As Management Professor Joe Kelly has said, meetings owe more to "theatre" than to "science." All of the actors who have played hard if not always well on the stage of your meetings want some response from the audience—even an audience of one. Make it a point, therefore, to compliment presenters who did a good job; show interest after the meeting in points raised by participants; follow up on questions asked in the meeting; and, simply make reference in day-to-day conversation about points made by participants. In this way, you will encourage meeting attendees to see the meeting not as the culmination (and termination) of group thinking but instead as the beginning of an action-oriented process.

Report on Meeting Results Up and Down the Chain of Command. You should not expect your superiors or subordinates to guess your conclusions about the meeting from the minutes they receive. In the form of a concise memo report, take time to sum up the major outcomes of the meeting for your superiors, perhaps on a For Your Information basis. Describe not only what the group decided but also interpret the group's thinking for upper management. What did this meeting mean? Where do participants think projects are going? Where do they want them to go? What problems do they foresee? Do you agree with the group? What's your view?

Depending on your management purposes, you may also want to send a report of sorts to your subordinates—but not the same report you sent upstairs. In the absence of some word from you regarding the meeting, your subordinates will believe whatever the company grapevine carries to them, seeds and all.

So. You've wrapped this meeting. Others are already telling you it's "the best meeting around here in years," "we actually got something accomplished," and "we finished on-time!" You're optimistic about a general reformation of meeting practices within your company. There's no reason that every meeting can't be as productive as this one, right?

To help you convert the masses, 3M Corporation publishes an excellent pamphlet, "Six Secrets to Holding a Successful Meeting," available by contacting "Six Secrets"/AV, Box 33600, 3M Center, St. Paul, MN 55133–3600.

Here's an extended excerpt that will also serve as a neat summary of our major concerns in this chapter:

Six Secrets to Holding a Successful Meeting

1. The right reasons: There are two very good reasons to have a meeting: the first is to share information and the second is to solve problems.

In general, information-sharing meetings should be held when you want to a) explain a plan or program, ensuring that everyone gets the same information in the same way; b) tell people what to do and how to do it—and how their job fits into the big picture; c) report what's been accomplished—and perhaps plan what's left to do; and d) win support or approval for an idea.

Problem-solving meetings should be held when you a) need to define or solve a problem by bringing together key players who may hold important pieces to the solution; or b) need consensus on a decision that has to be made.

In addition, there are two special situations when a meeting should be called: when you want to train people and when you want to build morale or recognize achievement.

2. The right people: The people you want in your meeting are those who need to know certain information or who can make a contribution. They should:

- ▲ have solid knowledge of the business at hand
- ▲ have the power to make a decision
- ▲ be responsible for implementing any decisions or carrying a project to the next step
- ▲ represent a group that will be affected by a decision made at the meeting

3. The right room: Every meeting room "speaks" to the people who enter it. Have you ever walked into a conference room and felt like you'd just stepped aboard a New York City subway at rush hour? Or been to a sales presentation in a ballroom where only nine customers showed up? Size is important. Large auditoriums tell people to sit back and listen, while small conference rooms ask for participation and teamwork. When a room is too small, people are uncomfortable and become unproductive. When a room is too large, they tend to sit far apart, inhibiting discussion.

4. The right equipment: There are three immutable laws you should observe when it comes to equipment. First, it has to be compatible with the kind of visuals you have (for instance, an overhead projector for transparencies, a slide projector for slides, an easel for storyboards). Second, it has to be there. Third, it has to work.

Get your equipment early, so you can rehearse your presentation on the equipment you'll be using. If necessary, ask your supplier to brief you on how it operates. Learn how to troubleshoot simple problems. For instance, request an overhead projector that has a built-in backup lamp, so you have a quick solution to a burned-out bulb. For a slide projector, have an extra projection lamp on hand and know the replacement procedure.

5. The right visuals: Research shows that people retain only 10 percent of what they hear, but 50 percent of what they hear and see. That makes a powerful case for the use of visuals at all of your meetings. Photographs, technical drawings, and cartoons help the audience grasp important details, relationships, and concepts. Copies of advertisements and articles lend credibility to a presentation (if they can be seen from the back of the room).

6. The Right Skills: As a meeting leader, you have three important responsibilities. One is to guide the participants through the meeting in a way that will accomplish its purpose. The second is to encourage participation. And the third is to keep the meeting on track—and on time. Here are some tips on how to do this:

- Before the meeting, plan and prepare the content, make up and distribute an agenda, and check out the room and the equipment.
- At the start of the meeting, review the agenda and objectives and explain what role the participants have in the meeting.
- To encourage participation, ask open-ended questions, reinforce statements that highlight objectives and use examples to encourage the group to think along similar lines.
- To maintain control, ignore off-target remarks, ask questions related to the task at hand, and restate relevant points of the agenda.
- At the end of the meeting, summarize, state conclusions, and outline actions to be taken as a result of the meeting.

ROLODEX RECAP

▲ small meetings are often more focussed and more productive than large meetings

▲ important one-on-one meetings should be scheduled, with a time, place, and working outline or agenda specified

ROLODEX RECAP

▲ staff meetings should not be overloaded with attendees who have only marginal concern with the business of the meeting

▲ after receiving input from key decision-makers, the agenda should be circulated several business days prior to the meeting

ROLODEX RECAP

▲ the agenda should specify time, place, attendees, purpose, and the items for discussion or action, with a suggested time allotment for dealing with each item

▲ attitudes toward meetings change with the site. For an especially creative meeting, consider off-site meeting locations, including nontraditional sites

ROLODEX RECAP

▲ seating patterns at meetings affect interaction. Choose a seating plan to support the purpose of the meeting

▲ follow up after a staff meeting by reviewing and distributing minutes, giving feedback to meeting participants, and reporting results up and down the chain of command

6

Large Meetings

LARGE meetings—from 25 participants to hundreds of conventioneers—are not plumped-up small meetings, as more than one corporate leader has learned to his or her chagrin.

"It seemed logical to me that a large meeting would just differ in scale from a small meeting. So I reserved a huge room instead of a small one. I paid $1200 to rent a sound system instead of borrowing a podium mike. I hired a big name from the lecture circuit instead of forcing one of my managers to make a presentation. And the meeting flopped." The tale of woe comes from a Pennsylvania executive editor who retired last year, not entirely by choice. He had taken upon himself the responsibility of planning the annual company meeting for about 300 sales and marketing reps. He had planned a big small meeting.

Large meetings differ from small meetings in the mind of the participants, the company, and the meeting leader. Participants come to a large meeting expecting to be entertained, inspired, and motivated. Those same people would have very different expectations of a small meeting, where they would come for information and intense discussion leading to participative decision-making.

The company, too, has different expectations for large meetings. It wants the large meeting to change the attitudes of participants more than their knowledge base. The company wants participants to feel more deeply and in new ways about their jobs. Small meetings, where nuts and bolts get spilled out on the table, often dampen employee feelings of loyalty, ambition, and team spirit. Companies hold large meetings in an effort to rally the troops.

Finally, meeting leaders look upon large meetings quite differently from small meetings. For one thing, nothing in the meeting from the order of speakers to the number of hors d'oeuvres is left to chance. "The best advice I can give anybody organizing a meeting is to plan and plan and plan ahead," says Michael Olenski, vice president of Corporate Media Communications, Inc., of Philadelphia. Clem Abik, marketing manager at Imperial Life Assurance Company concurs, calling lack of preparation "a recipe for chaos." Nor do leaders expect to play a major shaping role in the content of large meetings, as they might in the case of small meetings. Someone else delivers the keynote address, the luncheon speech, and even the closing remarks. The leader of a large meeting is content to play master of ceremonies, not mastermind.

A LARGE MEETING IN YOUR LAP

So it falls to you—primary responsibility for a large national meeting eight months hence. The good news is that you'll be a company hero after a great meeting. The bad news is that you'll never recover professionally from a bomb.

This chapter will tell you the seven most important things you should know in preparing, conducting, and cleaning up after a large meeting. I'm assuming, of course, that you want to do the planning yourself.

You don't? That's understandable—and more than 300 companies specializing in meeting planning are at your service. You may want these addresses in your Rolodex:

Meeting Planners International
3719 Roosevelt Blvd.
Middletown, OH 45044

Professional Convention Management Association
2027 First Avenue North
Birmingham, AL 89823

Society of Company Meeting Planners
2600 Garden Road, Suite 208
Monterey, CA 92930

At sometimes considerable cost, a meeting planner will do for your meetings what a C.P.A. does for your taxes—absolute wonders. For example, meet Walter Green, chairman of Harrison Conference Services, a company that runs conferences across the U.S., hosting more than 150,000 people a year in 6000 meetings. Green recognizes that "most managers have no formal training in how to plan and run meetings." After ascertaining the company's wishes for the meeting and budget range, Green's firm offers virtual turn-key meeting packages. All the company leaders and participants must do is show up.

Or picture your next large meeting as produced by Corporate Media Communications, Inc., of Philadelphia. Vice president Michael Olenski begins by trying to "climb into the head of your audience. You need to know who your audience is and what these people are like." Olenski's company recently put together a large meeting for Ore-Ida food products at a national brokers' meeting in Reno. Using a "Back to the Future" theme, the meeting took brokers back through highlights of Ore-Ida's stellar growth during the last 30 years. Clips from the "Back to the Future" movie, a solid gold DeLorean, and a look-alike mad scientist completed the illusion and brought audiences to their feet in applause. "We're really in the entertainment business," says Olenski. "But you should not have glitz for glitz's sake. If the message is not there, you have failed."

Then there's Leslie Buckland, acknowledged impresario of the big business meeting. Buckland charges upwards of $1 million for an all-out extravaganza. But that's precisely what some companies need and want. His shows/meetings attracted the roving eye of *People* magazine, especially the meeting he staged in Singapore "for a large computer company, complete with a Pucci fashion show and a troupe of both Broadway and native dancers. Speeches by executives were simultaneously translated into four languages."

Buckland's company in 1986 surpassed $20 million in sales, and has now staged meetings for State Farm, IBM, and Mobil. "They don't remember themes," Buckland says. "The trick is to do something that is memorable."

But you have something a bit less pricey in mind, and the company wants you in charge every step of the way. That was the fate in 1987 of John Crump, executive director of the 17,000-member National Bar Association (NBA). "There are no secrets to meeting planning," says Crump. "It's really a question of time and effort." Crump is living proof that exciting, well-organized meetings can be planned in-house. He keeps his finger on the pulse of each of the hundreds of details that make a large meeting a success. "I tell the hotels that I'm the last word," Crump says, "and that cuts down on surprises."

DETAILS, DETAILS, DETAILS

For those of you cut from the same cloth as John Crump, here are seven things you have to know as an effective meeting organizer.

1. Choose the right location: According to recent surveys published in *Sales & Marketing Management*, these factors matter (in order) most of all to company decision-makers:

- ▲ quality of guest rooms
- ▲ quality of service
- ▲ food
- ▲ on-site meeting staff and equipment
- ▲ rates

Surprisingly, the availability of on-site or nearby entertainment scored low, as did proximity to an airport. Companies seem primarily interested in a gracious professional environment for their meetings, with no surprises.

You can quickly scan the facilities of a particular city or region by obtaining a "Meeting Sites Guide" from a chamber of commerce or hotel association. The sample on page 102 is from the South Coast Area Hotel Meeting Sites Guide, which lists facilities in the Orange County, California, region.

The right location, of course, may turn out to be in your own backyard—even the the company headquarters building itself. What it lacks in pizzazz for attendees can be made up (with money saved from hotel costs) by lavish buffets and other extras.

Whether on-site or off-site, don't pack the room with participants. "Choose rooms that can handle 20 to 30 percent more than the expected audience," says Michigan State Professor Robert E. Lucas.

For a major off-site meeting, spend the money to visit the top contending locations yourself. Executive Vice President Alvin LeFaivre at Marriott suggests that you "call the sales managers and have them walk you through the properties. If your basic impression of the public rooms is not good, keep looking. Your intuition is a valid guide."

A guide form for location selection appears at the end of this chapter.

2. Cut the best deal. Whenever possible, work directly with the resort's or hotel's sales director. Mike Hausman, for example, is international sales director for Marriott Hotels. "I can contact one of our hotels, ask them when they need the business most, and cut the best possible deal," he says. Hausman points out that a company "can cut a great deal if [the company's] people will meet on weekends." Hotels tend to have more empty rooms at such times, allowing some companies to invite employee spouses at no extra charge.

Most large meeting sites will want you to leave a deposit of 25 percent or pay for the entire expense of the first night's rooms. If you're not ready to commit cash to the deal, recommends Marriott's Alvin LeFaivre, "ask the manager to put a

South Coast Area Hotel Meeting Sites Guide			Sleeping Rooms	Meeting Rooms	Theater Seating
Los Angeles Airport	Amfac Hotel-LAX	8601 Lincoln Blvd. 213/670-8111	750	15	800
	Holiday Inn Crowne Plaza	5985 W. Century Blvd. 213/642-7500	612	15	675
	Holiday Inn-LAX	9901 S. La Cienega Blvd. 213/649-5151	403	5	250
	Hyatt-LAX	6225 W. Century Blvd. 213/670-9000	597	33	1500
	Los Angeles Airport Hilton & Towers	5711 W. Century Blvd. 213/410-4000	1280	40	1600
	Los Angeles Airport Marriott	5855 W. Century Blvd. 213/641-5700	1012	34	1200
	Sheraton Plaza La Reina	6101 W. Century Blvd. 213/642-1111	800	96	1000
	Stouffer Concourse Hotel Los Angeles	5400 W. Century Blvd. 213/216-5858	750	22	1500
	Viscount Hotel LAX	9750 Airport Blvd. 213/645-4600	570	17	800
Manhattan Beach	Barnaby's Hotel	3501 Sepulveda Blvd. 213/545-8466	128	6	180
	Radisson Plaza Hotel & Golf Course	1400 Parkview Dr. 213/546-7511	400	17	750
	Residence Inn	1700 N. Sepulveda 213/546-7627	175	2	50
Newport Beach	Four Seasons Hotel	690 Newport Center Dr. 714/759-0808	296	10	650
	Hotel Meridien	4500 MacArthur Blvd. 714/476-2001	440	22	800
	Marriott Suites Newport Beach	500 Bayview Circle 714/854-4500	250	14	180
	Newport Beach Marriott Hotel & Tennis Club	900 Newport Center Dr. 714/640-4000	600	22	1000

South Coast Business, March 88, p. 10

'first option hold' on the space—if another group wants your date later, the hotel will call you before letting them have it."

For meetings with over 100 participants, consider sending a formal "RFP" (Request for Proposal) to the convention and visitors bureau in the cities or locales under consideration. Jim Eaglin, chairman of the National Association of Blacks in Criminal Justice, describes his organization's RFPs: "In order to select the most interesting and appealing site, we submit a Request for Proposal . . . which describes all of our hotel, transportation, entertainment, and other conference-related needs in detail, as well as some information about our history and membership."

"In addition," says Eaglin, "the RFP lists specific items that we require from them, including each hotel's affirmative action plans, a statement on how the minority business community is likely to benefit from our conference, and an indication as to the kind of receptivity we can expect from local political and community leadership."

In such RFPs, many meeting planners forget to tell the convention and visitors bureau about the content of their gathering—its theme, proposed outcomes, and major events. Laverne Morrow of Emprise Designs, Inc., a St. Louis-based meeting planning firm, says that "most groups are really good at negotiating with hotels and airlines, but they fall short in structuring the content, topics, and activities of their meeting."

And don't be shy about asking for extras. Convention and visitors bureaus can often arrange a welcoming banner, a proclamation from local government, dignitaries, ceremonial bands, shuttle service, and event sponsorship. When MPI met recently in Philadelphia, the mayor declared "MPI Day" and ordered a welcoming banner hung from city hall. Little things mean a lot!

3. Get the deal in writing. There's a special place in hell for the hotel employee who forgets to write down a company's reservation, or double-books large meetings. John Crump at the National Bar Association goes so far as to send confirming letters for reservations "not just to the convention services manager and sales manager, but to everyone in the hotel's hierarchy." Radisson Hotels has formalized the confirmation process with its Assured Meeting Agreement specifying all aspects of the meeting plans, including room setup. The agreement provides for cash rebates to the company in the event that Radisson fails to meet the precise terms of the agreement.

Mike Hausman of Marriott Hotels points out that his industry is notorious for turnover. The hotel employee you deal with on January 1 may have moved on by the time of your meeting. "A handshake commitment may not mean anything to the staff in place when you arrive," says Hausman.

You'll also want to get a written commitment from any site that plans to host other major meetings at the same time as your meeting. Marriott's LeFaivre warns that "with a class reunion on one side of you and a gathering of professional comedians on the other, your strategy-development meeting can be hampered. Find out what other groups are booked for the same times your group is."

4. Learn and use the language of meeting setups and equipment: Most snafus occur because company representatives have misused or misunderstood basic terms for room arrangements and equipment.

Lacking a shared language with the company, hotel meeting planners must be mind-readers or detectives to arrange rooms appropriately. Scott Picker, director of conference services at the Westin South Coast Plaza in Orange County, California, assembles clues one by one from hints dropped by company representatives. "I ask, for example, if people at the meeting will be using notebooks," he says. "If the answer is yes, then I say, 'O.K., you want a schoolroom setup.' " Company folk particularly shy away from discussing audiovisual equipment, Picker says. "They say, 'This is so technical,' but it's not." At the end of this chapter are handy AV and set-up terms, with definitions, that can help you communicate exactly what you want from hotel representatives.

Planning Form for American Cancer Society Meetings

Before the meeting	Target Date	Date Completed
Speakers or participants		
___ Written acceptance	___	___
___ Written agreement on honorarium	___	___
___ Orientation to A.C.S. program	___	___
___ List of any special equipment needed	___	___
___ Advance information or material	___	___
General meeting arrangements		
___ Site of meeting	___	___
___ Written confirmation of reservation	___	___
___ Written agreement on cost of meeting room(s)	___	___
___ Written agreement on cost of meals and menus	___	___
___ Decision and agreement on meeting facilities required	___	___
Promotion and publicity		
___ Biographical data on speakers	___	___
___ Photos of same	___	___
___ Advance copies of speeches or papers	___	___
___ Announcement of meeting	___	___
___ Contacts with reporters	___	___
___ Preparation of posters or flyers	___	___
___ Distribution of same	___	___
___ Advance mailings	___	___
___ Phone followup	___	___
___ Preparation of printed program	___	___
___ Distribution of same	___	___
___ Secure and instruct photographer	___	___
Meeting facilities and room arrangements		
___ Seating arrangements	___	___
___ Speakers' table arrangements. Check on—	___	___
___ Location of electrical outlets	___	___
___ Location of P.A. controls	___	___
___ Where special equipment will be placed	___	___
___ Arrangements for press	___	___
___ Facilities for social activities	___	___
General		
___ Prepare registration cards	___	___
___ Prepare badges or name tags	___	___
___ Prepare lists of complimentary and prepaid reservations	___	___
___ Prepare lists of materials, supplies and equipment to be taken to the meeting	___	___
Time of meeting		
___ Complete registration setup	___	___
___ Set up identifying sign or poster	___	___
___ Final check on arrangements and facilities	___	___
___ Final check on special equipment	___	___
___ Place cards at head table	___	___
___ Set up exhibits or displays	___	___
___ Distribute and collect evaluation forms	___	___
After the meeting		
___ Tips to hotel staff	___	___
___ Return borrowed or rented equipment	___	___
___ Followup news story	___	___
___ Thank you letters	___	___
___ Prepare and distribute minutes or proceedings	___	___
___ Pay bills	___	___
___ Summarize evaluation forms	___	___
___ Report to Board of Directors	___	___

"And, gentlemen, this trade show is really going to be different. We will not allow a single talking robot into the exhibit area!"

"If there's one place where things go wrong more consistently than anywhere else, it's with slide shows, films, and tape machines," says Joe Will, public relations director of Trimac Ltd. Knowing what to check can go a long way toward solving this problem. Prior to your next meeting, check off each of these items compiled by Homer Smith, contributing editor of *Sales & Marketing Management*:

1. On arrival, check with site personnel for agreement on list of required AV equipment, schedule, and availability.
2. Check the day before the meeting session on arrival of required AV equipment.
3. Check that equipment is the kind requested. Follow up on missing units.

4. At least one hour before the session, check on AV readiness.
5. Check out projection and sound equipment to familiarize with use and to see if it is working.
6. Load first AV element to be used. Check projector operation, focus, screen size and placement, sound level, film or slide advance. Is screen large enough for group and properly positioned?
7. Are power cords taped or protected against tripping and accidental disconnection?
8. Become familiar with corrective actions if needed: replacement of projection and exciter bulbs, fuses, critical connections; loss of film loop; sound level; jammed slides.
9. If speaker will operate slide projector, is the remote control device working?
10. Is it taped to prevent tripping?
11. Do you know how to control the room lights and darken windows?
12. Who will control lights and darken windows when required?
13. Are there lights at the lectern so speaker can read his or her script? Do the bulbs light?
14. Are there chalk and eraser at the chalkboard? Is the chalkboard clean?
15. Is it large enough, stable, with a good writing surface?
16. Are there crayons or workable felt markers at the chartpad? Is there sufficient paper on the pad? Should there be more than one pad?
17. Check inventory of AV software: do you have the right film, slides, tapes, handout materials, etc.?
18. Is AV software in the correct starting position? Examples: slides in right sequence, right position; tapes at proper spot to open; overhead transparencies in proper sequence.
19. Is overhead projector in convenient position for speaker? Is screen positioned so speaker and projector do not block the view? Has screen been adjusted to minimize keystone effect?

20. Are there proper markers or crayons with the overhead projector for the speaker to use? Are there blank films or a film roll available if needed?
21. Where will AV equipment be stored when not in use? Is it safe?

5. Find out what your consultants and speakers need. You have your act together—but are you also prepared to help your consultants and speakers get their show on the road? Mike Blewitt, manager at Arrowwood of Westchester (N.Y.) conference center, reminds meeting planners to consider the needs of outside people participating in the meeting. "The consulting business is booming," he says, "and much sales training is being done by consultants. Few meeting planners check with consultants to find out their AV needs."

6. Think through the type and timing of meals. In *Sales and Marketing Management* surveys, the quality of food provided ranked consistently as one of the top three factors of what matter most to meeting participants. MIT researcher and food specialist Dr. Judith Wurtman finds that "people plan everything about meetings; everything but the right food." Food has a direct bearing upon the productivity of a meeting because, in Wurtman's words, "there is a definite, measurable relationship between what one eats and how one feels and performs." She recommends less food, but better food: "Conferees will feel more energetic, more attentive, and more motivated if you underfeed them." Serve foods high in protein and fiber, but low in fat—as in the following suggested menu from the New York Hilton.

When releasing meeting participants for meals at hotel restaurants, plan ahead to miss the rush hours. "It's wise not to eat at 12:00 or 12:30," says Arrowwood's conference manager, Mike Blewitt. "Everybody wants to eat then, and you don't get great service." Dinner is better scheduled at 6:00 or 8:00, he says, than the prime hour of 7:00 or 7:30.

A final consideration: don't bog down your meetings by the clink of plates, cups, and silverware. James Kimes, a vice

Buffet or Reception

Choose some or all of the following:

Crudités with cottage cheese or yogurt-based dip

Beef, chicken, and seafood kabobs.

Smoked salmon, smoked oysters, smoked duck.

Boiled shrimp.

Marinated herring, marinated mussels.

Marinated mushrooms, marinated artichokes.

Miniature tortillas filled with diced chicken.

Steamed Chinese dumplings.

Blintzes filled with ricotta cheese and topped with caviar.

Fresh fruit mélange.

Fruit kabobs, made with strawberries, pineapple, melon balls

Lunch

Gazpacho Andaluz

Salad Niçoise, made with chunks of tuna, whole blanched string beans, and sliced fresh mushrooms on a bed of lettuce; garnished with tomato wedges, sliced fresh yellow squash, blanched broccoli buds, and one sliced new potato.

Dijon vinaigrette dressing.

Whole wheat and pumpernickel raisin rolls.

Sliced fresh fruit in season.

Coffee, tea, or decaffeinated beverage, as desired.

Afternoon Break	
Frozen yogurt in cups. Trail mix (made with dried fruit, coconut, nuts).	OR Barbecued chicken chunks, smoked oysters, boiled shrimp.
Fresh fruit, such as pineapple, strawberries, melon.	Raw vegetable crudités with yogurt dip.
	Zucchini bread, cranberry bread. Popcorn, no-salt pretzels.
Coffee, tea, (both caffeinated and caffeine-free), fruit juices, mineral water.	Coffee, tea, (both caffeinated and caffeine-free), fruit juices, mineral water.

clink of plates, cups, and silverware. James Kimes, a vice president at Vestal Laboratories, advises "keep the 'goodies' out of arm's reach until break time. Nothing is more demoralizing to someone trying to get a point across than to have half of the people pouring coffee and the other half munching Danish."

Kimes tells the amusing story of attending a board meeting where it was decided that the group would eat during the presentation of his report. Somewhat disconcerted, Kimes began his presentation by telling his feeding audience that he now knew what it must feel like to do a scene from "Clan of the Cave Bear." Meals were not served at board meetings thereafter.

The question of whether or when to serve alcohol during large meetings depends very much upon a company's own corporate culture. According to Suzanne Hall in *Successful Meetings* (Aug. 1988), "planners and food and beverage directors almost unanimously report that, at least during the day, alcohol is definitely out." Hall quotes food expert Judith Wurtz: "The brain-dulling potential of alcohol is so great, so hazardous to the success of your working meal, that I do not even want to suggest a safe upper limit. There is none."

Also to be considered is the risk or temptation posed to alcohol abusers or recovering alcoholics. Richard Frances, chairman of the American Psychiatric Association's committee on alcoholism, grants that alcohol is often used "as a social lubricant—a way of making it easier for some people to get in touch with their feelings or to take risks." But, says Frances, 'for recovering alcoholics who go to meetings, it can almost be torturous to be away from home, away from family, away from support systems, away from their Alcoholics Anonymous groups. . . . I've had a number of patients who were sober for long stretches, only to slip and start drinking again while away at meetings."

For employees who think they're drinking on the company's tab at meetings, there may be rude surprises. A 1988 survey by the Administrative Management Association showed that only about a third of companies will reimburse their employees for drinks at business meetings, down from two-thirds of companies in 1980.

7. You may want to start your meeting during travel. Let's say that you have 100 meeting participants flying to your meeting, at an average roundtrip fare of $300. That's $30,000! For that kind of money or less, you may be able to arrange quite special transportation that supports your meeting purpose and theme. Sentimental Journeys Aviation of Oakland, California, for example, has recreated the mystique of DC-3 air adventures in the 1940's, complete with stewardesses in period uniforms. They fly to all major resort and urban areas in California, with a base rate of $1500 per air hour. For information, call (415) 667-3800.

Or for the jet set, there's MGM Grand Air—the ultimate way to fly L.A. to New York. Luxuriously appointed 727 jets include staterooms, leather swivel seats, large-screen TVs, and a sumptuous bar/buffet. The cost is $5,500 per air hour, including the services of five flight attendants. Call (800) 422-1101.

Your connection for other flight possibilities may well be Flight Time. The company can provide virtually any kind of commercial aircraft, from Lear jets to helicopters to Concordes and 747s, and will fly wherever you wish. By chartering your own plane, you may save money and capture the attendance of employees on board for a few more precious hours of meeting time. Call (800) 445-0404.

For the John Maddens of business travel who hate to fly, there's Amtrak's Special Movements Bureau—a subagency specializing in moving large groups in style. In 1988, the SMB arranged a special train for the 1,200 member New Jersey Chamber of Commerce and another for a Florida law enforcement group. "Last year Amtrak made $5 million in Special Movements revenue," says SMB manager Ladislav Shrbeny, "indicating that we do this for a living and do it successfully." Call (800) USA-RAIL.

If you're moving a smaller group, you can attach a very special private coach of your own to an Amtrak train. One of the more spectacular of these is described in *Successful Meetings* as "the opulent 'Virginia City,' . . . a circa 1928 'tasteful bordello' that features a pink dining room with a crystal chandelier and a lounge with a marble gas fireplace." Such cars can accommodate 12 to 15 day passengers and sleep 6 to 8. For information about rates and availability, obtain the Charter Referral Directory from the American Association of Private Railroad Car Owners, 224 Orr Drive, Somerville, New Jersey 08876.

Some other creative possibilities: Mississippi River steamboats are available for both transportation and meetings. One company specializing in such services is the Delta Queen Steamboat Company [(800) 543-1949]. Special European-style tour buses (with optional jazz band or tuxedoed serving staff!) are available through Keeshin Charter Service [(312) 254-6400] and Conventions à la Carte [(504) 329-2121]. But in making any travel arrangement, warns Wayne Smith, executive director of the United Bus Owners of America, make sure you are dealing with "reputable, licensed, insured transportation movers." Your company may be held liable in a third-party legal action.

MAKING YOUR LARGE MEETING SPECIAL

So the chairs are set up, the name tags are printed, and the shrimp's on the barby. Now how can you add to the meeting that extra something that translates into improved morale and performance?

First, don't wait for the sessions to begin. Contact participants in advance to get them thinking about the issues, themes, and general excitement of the coming meeting. A pharmaceutical company planning a large meeting in Hawaii sent a series of "teaser" letters to participants. One letter purported to come from a Hawaiian and invited participants to a luau. Another brought them up to date on weather and fishing information. The company's goal was to raise expectations—to have their participants half sold on the meeting before they even arrived.

Breakfast in bed delivered to each participant's room, says Helen Moskovitz, president of Meeting Services and Convention Consultants, can start things off on a high note. "It gets their interest up," she says, "and they're more apt to receive information at the meeting."

Or you can remind your upcoming group what a great time a previous group had at your meeting. One company sent out a videotaped selection of the previous year's meeting (including the fun) to invitees of an upcoming meeting. The message was clear: here's what's possible—now let's outdo them!

Second, plan for a reasonable amount of what meeting managers call "creative downtime"—time for participants to relax, network, digest (both physically and mentally), and sightsee. If you suspect that social interaction could use a boost during breaks, ask your hotel for imaginative suggestions. At the Vista International Hotel in New York, for example, banquet sales manager Nicholas Calderazzo puts on refreshment breaks titled "Chocolate to Die For." There's everything for the good, bad, and pudgy: hot fudge sundaes, truffles, brownies, chocolate chips, and crushed Oreos. The Omni International Hotel has a choc-attack break as well. Chefs sculpt varieties of chocolate truffles into a five-foot tree of sorts. Some attendees call this feature "MBS"—Management by Sugar.

Third, translate the theme of your meeting into visual form—something that hits attendees between the eyes. The Sales Association of the Chemistry Industry—not necessarily wild and crazy guys—set up a boxing ring right in the middle of their country club meeting room. The International Association of Conference Centers divided attendees into teams and assigned each a case study involving several scavenger hunts and the making of a videotape for presentation to the group. Just at the edge of the far side are Pastiche, Inc., and Chez-Zam, both New York purveyors of . . . well, the bizarre for meetings. Pastiche hires out mimes and actors who play the part of larger-than-life statues or objects. A mime done up as a nine-foot Christmas tree goes about embracing meeting participants. A pseudo-statue of Moses suddenly springs to life to lead attendees to the dining room. Chez-Zam features pop art attention-getters: human-sized shoes, gangly octopi, and walking tennis rackets.

Expensive foolishness? Expensive, yes. Foolishness, no. Any meeting planner who thinks the most memorable aspects of the meeting come from speeches and handouts belongs back in Group Dynamics 101. As Professor of Management Joe Kelly points out in *Personnel,* "running a meeting is an exciting activity. . . . When you consider the manager who actually has to run the meeting, you soon realize the need for myth, magic, and meaning. Producing and directing meetings become the art of transforming inputs into outputs with the addition of value. This 'two plus two equals five' aspect of meetings demands a certain savoir faire that owes more to theatre than it does to science."

THE BOTTOM LINE

So many people have had a hand in shaping your large meeting—a company committee, the hotel advisors, convention and visitors bureau staff, special effects consultants, menu planners, and others. When things go right, they're all glad to take the credit. (In fact, you'll probably see your company mentioned in their next brochure.)

But when things go wrong? Nancy N. Nolan, senior meeting planner at Marriott Hotels, tells it like it is: "The manager responsible for arranging the meeting should be well aware of every nuance and detail instrumental in guaranteeing the success of the meeting—each rough edge will reflect on the host company and, ultimately, the person responsible for arranging the event."

You can prepare yourself further for planning large meetings by subscribing to the following magazines:

Successful Meetings
633 Third Avenue
New York, NY 10164–0735

Meetings & Conventions
Murdoch Magazines
1 Park Avenue
New York, NY 10016

The Meeting Planner
3719 Roosevelt Boulevard
Middletown, OH 45044

Meetings and Incentive Travel
1450 Don Mills Road
Don Mills, Ontario, Canada M3D2X7

Corporate Meetings and Incentives
757 Third Avenue
New York, NY 10017

A book, *The Convention Liaison Manual*, is available from the Convention Liaison Council, Box 1738, Alexandria, VA 22320.

Also highly recommended are the following forms and checklists, reprinted by permission of Marriott Hotels.

Hotel Booking Checklist

Hotel: ______________________________

Address: ______________________________

Telephone: ______________ Telex: ______________

Contact: ______________ Title: ______________

Company: ______________________________

Address: ______________________________

Telephone: ______________________________

Number of rooms in hotel _____

singles _______ doubles _______ twins _______

twin doubles _______ queens _______ kings _______

1-bedroom suites _______ 2-bedroom suites _______

parlors _______ cabanas _______

Check-in time _______ Checkout time _______

Room Tax _______

Complimentary rooms offered ______________

Deadline for receiving room list ______________

Billing policy:

Master billing available _____ yes _____ no

Special billing arrangements

Rooms billed to company _____ yes _____ no

Rooms billed to individual _____ yes _____ no

Incidentals billed to individual, rooms to company _____ yes _____ no

Credit cards accepted ______________________

Type of property city hotel _____ airport hotel _____

suburban hotel _____ resort hotel _____

conference center _____ country lodge _____ ship _____

Distance in miles and time from airport ______________

Hotel/airport shuttle service yes _____ no _____

Complimentary yes _____ no _____

Parking: On-site ___ Complimentary ___ yes ___ no ___
Fee $ ___________
Off-site ___ Complimentary ___ yes ___ no ___
Fee $ ___________
Valet yes ___ no ___

Other groups meeting in hotel at time of your meeting______

____

How activities of other meeting will affect your meeting ___

____

Meal guarantee policy _____ hours in advance
_____ % set up over guarantee

Special security provisions for attendees _______________

Special storage arrangements for on-site materials _______

____

Hotel is sending brochure of space to you _______________

Hotel is sending sample menus and prices for breakfast, lunch, coffee breaks, receptions, and dinners as price guidelines _______________________________________

____

Tax on food functions _________________________

Service charge policy on food functions _______________

Meeting Profile

Company/Association Name: ______________________

Address: ______________________

Telephone: ______________________

Name or Description of Meeting: ______________________

Number of Attendees: ______________________

Meeting Objectives: ______________________

Preferred Type of Property:

☐ City Hotel ☐ Airport Hotel

☐ Suburban Hotel ☐ Resort Hotel

☐ Conference Center ☐ Country Lodge

☐ Ship

Preferred Location:

USA (☐ NE ☐ SE ☐ Midwest ☐ NW ☐ SW ☐ Hawaii)

International (Country): ______________________

Preferred Climate: ______________________

Preferred Quality:

☐ Economy ☐ Moderate ☐ Deluxe ☐ Super Deluxe

Importance of Accessibility:

☐ Airport ☐ Train ☐ Public Bus

☐ Taxis ☐ Rental Cars ☐ Private Cars

☐ Parking

Importance of Amenities:

☐ Restaurants ☐ Entertainment ☐ Shopping

☐ Sightseeing

Recreational Needs:

☐ Golf ☐ Tennis ☐ Swimming ☐ Health Club

Special Amenities Required: ______________________

Length of Meeting (# of days): ______________________

Preferred dates:
First choice ____
Second choice ______________________________________
Third choice ____

Suggested Properties (if known):
1. ____
2. ____
3. ____

Accommodations Schedule

Kind of Room	**# Needed**	**Primary Arrival Date**	**Primary Departure Date**
Single			
Double			
Twin			
Twin Double			
Queen			
King			
Parlor—small or large			
1-bedroom suite			
2-bedroom suite			
Cabana			

Daily Function Schedule

Date ________________ Day ________________

Time	Function	# of People	Room Setup
	Breakfast		
	Meeting		
	A.M. Break		
	Meeting		
	Luncheon		
	Meeting		
	P.M. Break		
	Meeting		
	Reception		
	Dinner Dance		
	Entertain-ment		

Meeting Room Checklist

Determine the best setup for each of your functions.

____

____

____

Are the meeting room sizes appropriate for your group size and proposed setups? ____________________

What obstructions must you plan for? ______________

What are your requirements for podiums, platforms, audiovisual equipment, and other special equipment? __________

____

What special arrangements must be made for signs/decorations? ____________________________

____

Determine the availability, source and cost of each of the following. If a price list is available, get it.

Microphones ____________

16-mm Film Projector ____________

Remote Control ____________

Computer/High Tech Equipment ________

Lectern ____________

Flip Chart ____________

Electric Pointer __________

Tables and Skirting _______

Overhead Projectors _______

Slide Projector and Carousel __________

Videotape Playback Equipment __________

Tape Recorders __________

Screen ____________

Blackboard ____________

Chairs ____________

Risers ____________

Supplies (pads, pencils, ashtrays, matches, water, glasses, telephones)

____

Wastebaskets ____________

Outlets ____________

Coat Rack____________

Lighting ____________

Temperature Controls_____

Floral Arrangements ______

Handicap Considerations ____________________

Using AV Terms with Precision (or, Getting What You Asked for)

The purpose of the following list of terms is not to define the obvious, but instead to set down standard names for audiovisual devices, with occasional comments on less common machines. Use the list to check your own use of terms and to throw a wider net when considering what specific AV technologies you can use in your meetings.

Acetates. Plastic "transparencies" for use with an overhead projector. Come in single sheets, or in rolls that can be scrolled across the light table of projectors equipped for same.

Audiovisual teleconference. Meeting participants appear on large projection screens, usually with full motion and high fidelity sound. See Chapter 8.

Beta videotape. The less common form of videotape—VHS is more popular by far. Beta videotape will run only on Beta video cassette players.

Blackboard. Useless without chalk and erasers. Consider a variety of chalk colors.

Cable TV. May prove to be a valuable and inexpensive vehicle for your larger business productions. Many stations make airtime and production facilities available to business for public interest programming.

Closed circuit TV. "In-house" television with signal relayed by means of hard-wired circuitry as opposed to broadcast on public airwaves.

Compact disc player. Extremely high fidelity sound from an optical disc "read" by laser.

Computer-assisted instruction (CAI). Programs that make use of on-screen text, graphics, and sound for instruction/training. Most CAI uses *"branching" programs that adjust to the learner's pace and progress.*

Computer projector. A projection device that projects monitor images from the computer onto a movie-sized screen.

Computer storyboard. A program of text, graphics, and sound stored on disk and delivered by computer monitor.

Filmloop. A relatively brief (30 second to 3 minute) film production, usually demonstrating a product or process.

Filmstrip. A slideshow, with individual frames attached into a single strip of film. The order of frames cannot be changed.

Flip chart. A large newsprint (or better quality) paper pad affixed to an easel. Pages can be prepared in advance or on the spot during a presentation or discussion. Provide broad-tipped markers in a variety of colors.

Lapel mike. A microphone that attaches to clothing. May be connected to an amplifier by wire or, preferably, by a wireless signal sender.

Lighted lectern. A must when the speaker needs to refer to notes during a media presentation in a darkened room.

Movie projector. Usually 16 mm for professionally produced films. Make sure someone can load the projector—"automatic loading" isn't as simple as it sounds.

Opaque projector. A large optical projector for projecting non-translucent originals onto a screen. Try it before your meeting—opaque projectors have limited clarity, distance of projection, and duration of use. Originals tend to wrinkle or burn when projected for more than a minute.

Overhead projector. For use in projecting transparencies (acetates). Notorious for burned-out projec-

tion bulbs—bring a spare. Older models have noisy fans that may distract audience or obscure speaker's words.

PA. Public address system, usually in the form of a central microphone with signal transmission to a number of speakers distributed throughout the building. Quality of transmission may not be suitable for more subtle sounds, including music.

Posters. Popular particularly at conventions where a product story must be told quickly to passersby. Provide large poster-display partitions.

Reproducing greaseboard. A large greaseboard that will generate photocopies of anything written on it. Extremely useful for brainstorming and planning sessions.

Room (or table) mike. A microphone that will pick up sounds from an entire room, not just from a single speaker.

Slide projector. Make sure you specify carousel or box slide trays.

Transparencies. Plastic (acetate) sheets or rolls for use with overhead projectors.

U-Matic. Videotape somewhat larger than 1/2'' VHS or Beta tape.

VCR. Videocassette recorder

VHS. The most popular format of videotape.

Videocassette recorder. Also known as VCRs, these recorders come with VHS, Beta, or U-Matic capability.

Videodisc. Monitor or on-screen projection, including sound, from an optical disk ''read'' by laser.

Room Set-Up Terms

Like the list of AV terms, this list is intended to help you consider a wide range of alternatives when planning seating arrangements—and to name correctly the pattern you eventually choose.

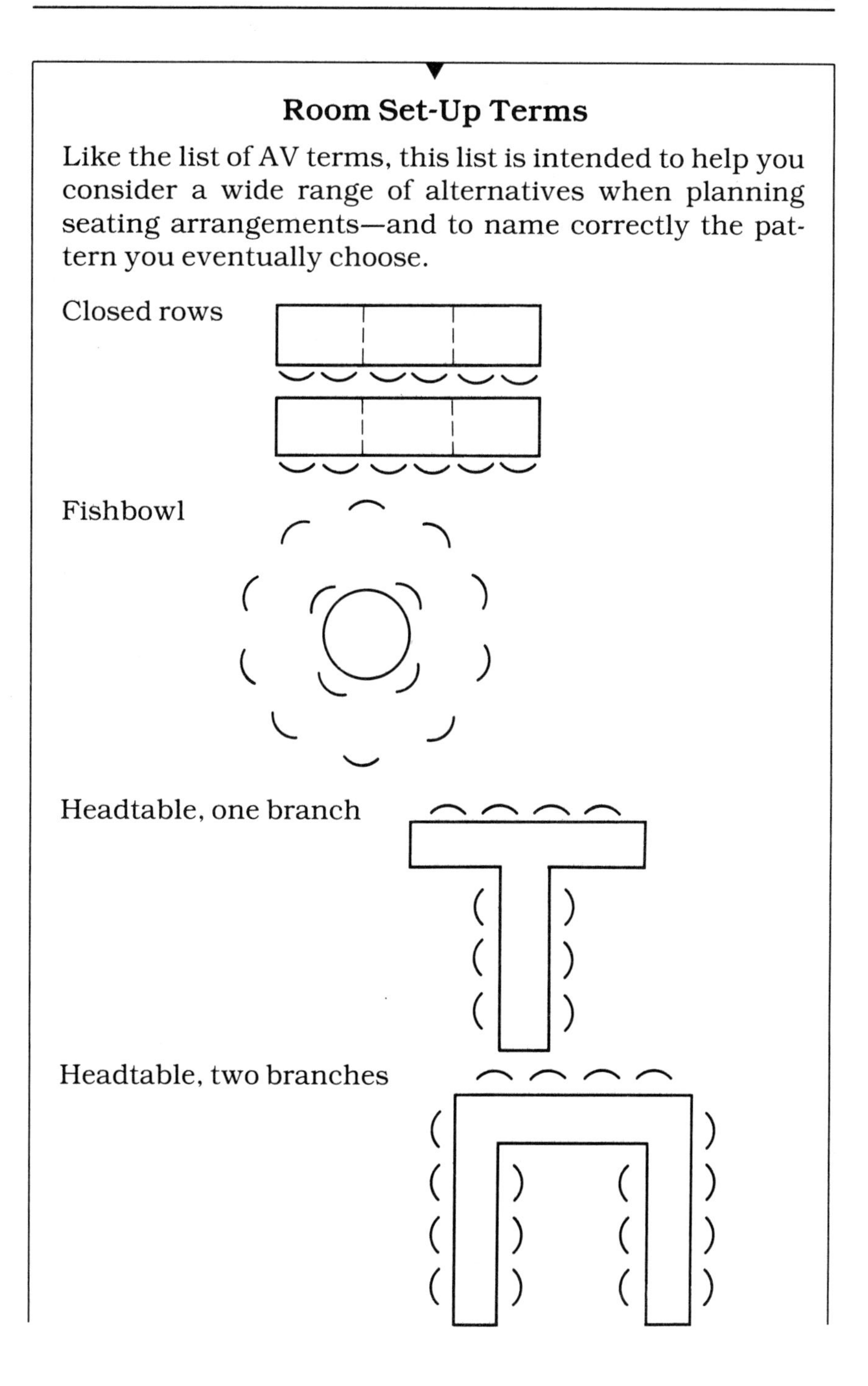

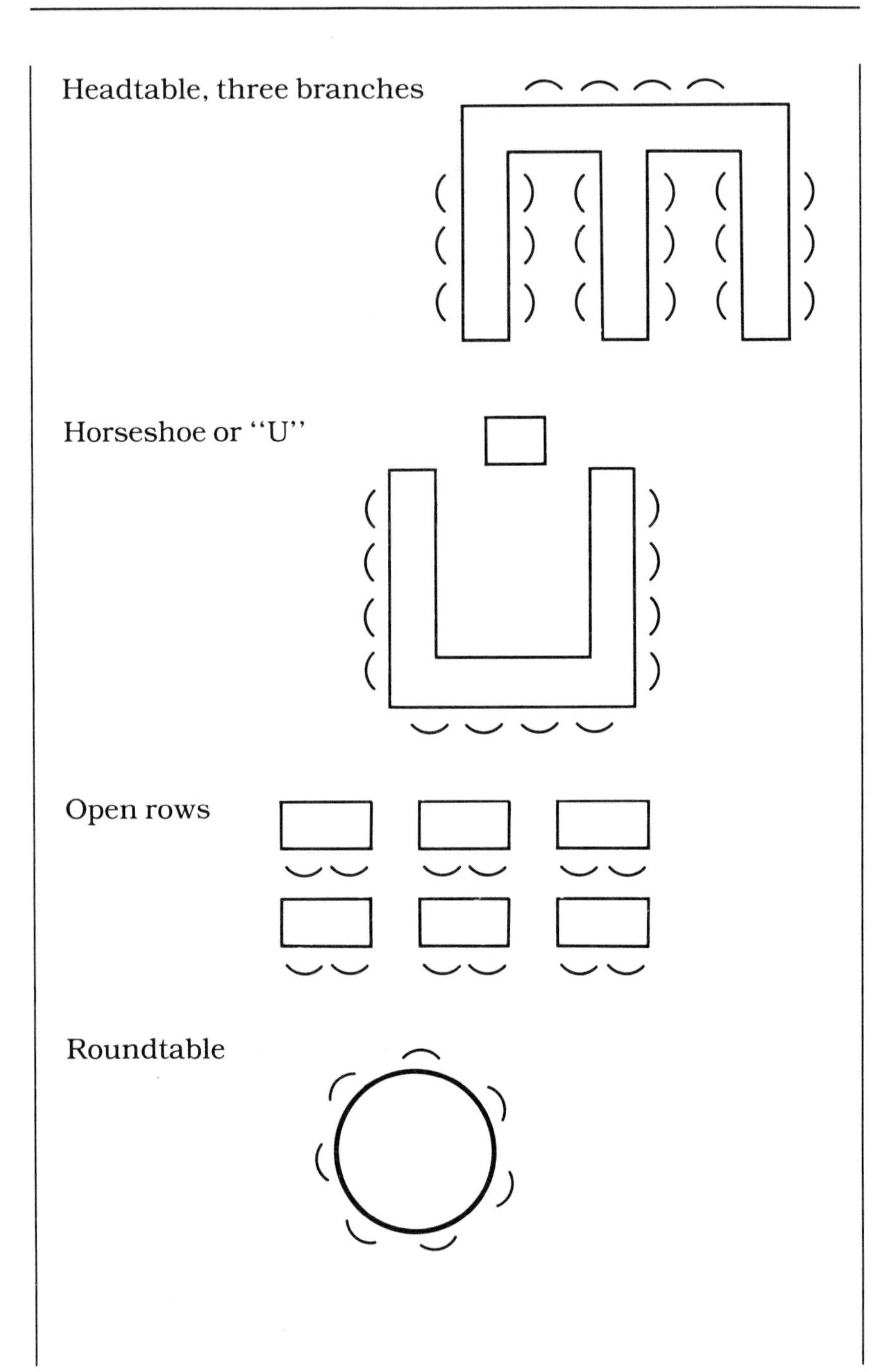
Headtable, three branches
Horseshoe or "U"
Open rows
Roundtable

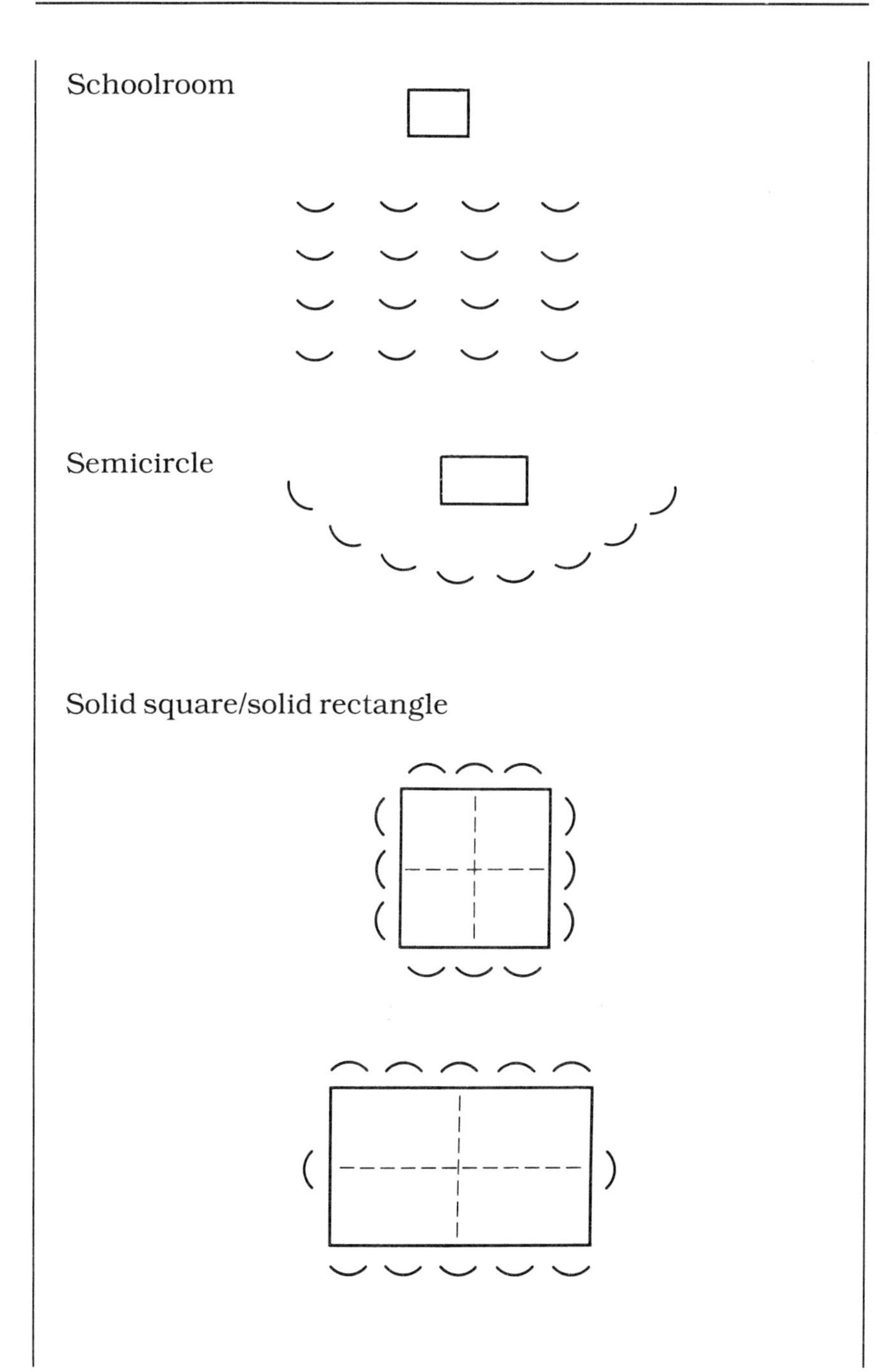
Schoolroom
Semicircle
Solid square/solid rectangle

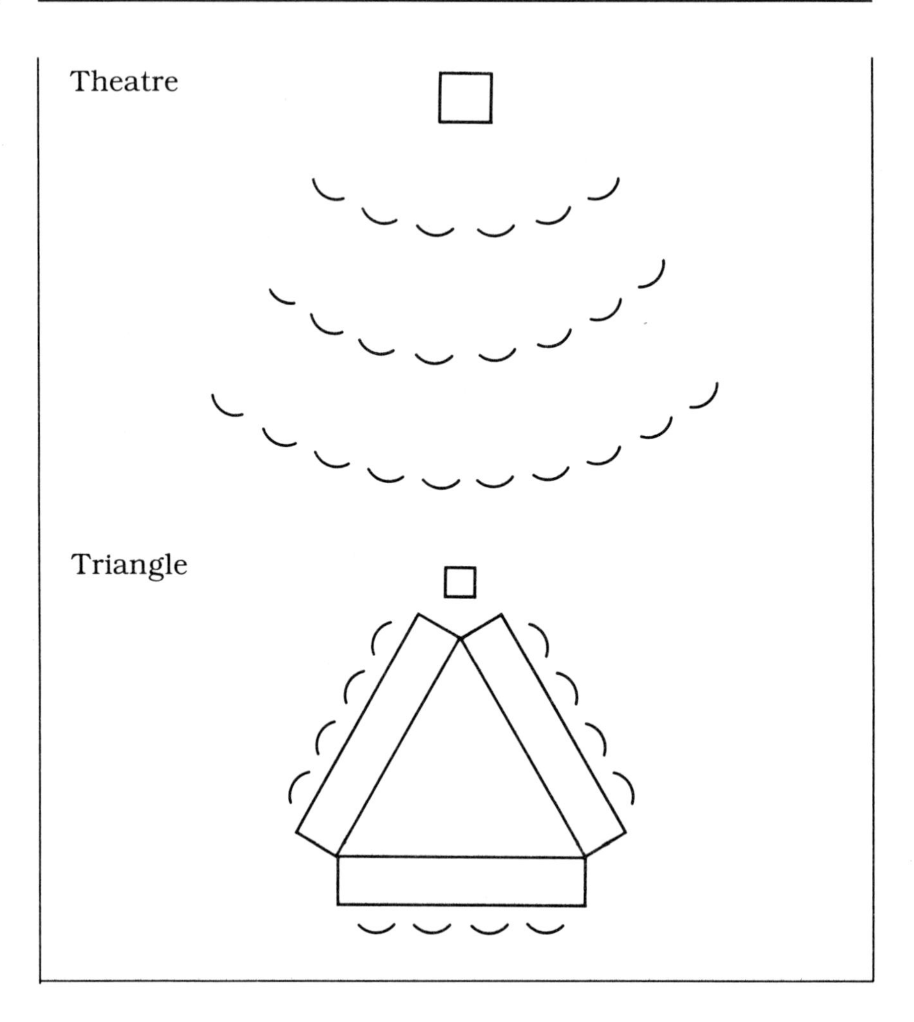
Theatre
Triangle

FILMS AND VIDEOTAPES ON MEETING SKILLS

The following are available in VHS, U-Matic, and 16 mm formats. Check for Beta availability.

Can We Please Have That The Right Way Round?, available through Video Arts, Los Angeles, CA. 22 min.

Communicating Non-defensively, available through McGraw-Hill Films. 21 min.

Face to Face Payoff: Dynamics of Interviewing, available through Video Software and Production, Yonkers, NY. 40 min.

Group Dynamics: Group Think, available through McGraw-Hill Films. 22 min.

How to Conduct Meetings, available through Time-Life Multimedia. 25 min.

Listening: A Key to Problem Solving, available through AIMS Media, Glendale, CA. 35 min.

Making Your Case, available through Video Arts, Los Angeles, CA. 27 min.

Management: Listening for Understanding, available through University of Mid-America. 35 min.

Meeting in Progress, available through Barr Films, Pasadena, CA. 40 min.

Meetings, Bloody Meetings, available through Video Arts, Los Angeles, CA. 35 min.

More Bloody Meetings, available through Video Arts, Los Angeles, CA. 40 min.

Presentation Excellence, available through CBS Fox, Farmington Hills, MI. 25 min.

Speaking Effectively to One or One Thousand, available through Sunset Films/Toastmasters International. 21 min.

The Anatomy of a Presentation, available through Roundtable Films, Beverly Hills, CA. 30 min.

7

Meetings in the 1990s: Issues and Options

AS we move toward the end of the century, new doors seem to be opening on all sides in the management of corporate meetings. We're leaving behind "Theory X" meetings—a rigid, boss-talks/employees-listen approach—in favor of "Theory Y" meetings emphasizing group discussion and decision-making. For better or worse, we seem to be meeting more often. And, as suggested in previous chapters, we're meeting more creatively in new locations and with imaginative flair.

Although crystal balls tend to fog up just when you need them most, I'll venture a prediction of four important "growth areas" for meetings in the 1990s:

1. MAKE ROOM FOR THE KIDS

So Mom's invited to her annual company meeting, and this time Dad is going along too, as loyal corporate spouse. If, that is, a babysitter can be found . . . for two kids . . . and a dog . . . for four days. Companies are recognizing that parents often find it easier (and sometimes more fun) to bring the kids along to big meetings, particularly when those gatherings are held at resort locations. More than 10 percent of business travel in 1988 included children, according to Julie Bowers at the U.S. Travel Data Center: "That accounts for nearly 206 million trips annually," up substantially from previous years.

Many families combine business travel with a vacation—three days on business in Southern California, for example, and the rest of the week at Disneyland and the beach. Companies aren't fighting the feeling. "I now run between two-and-four kids' programs a month for meeting groups," says Eileen Paris, activity supervisor at Marriott's Orlando World Center.

Why are even the most conservative companies welcoming the brat pack along to major meetings? First, it costs companies very little. Room accommodations with "children free" are relatively easy to negotiate at resorts eager for the kids' appetites at restaurants and snack bars. Airlines continue to offer reduced family fares. Second, many corporations find that employees who bring along the family tend to form close-knit bonds more quickly with other employees—the elusive "corporate family" so often espoused and so rarely realized. When families mix with families at corporate affairs, it appears that employee turnover goes down and morale increases. Finally, attendance at meetings improves. Employees don't have to arrive late and leave early due to family responsibilities. In long distance calls alone, many companies achieve a net savings by inviting families to conferences.

The employees benefit as well. "Are you really doing your top incentive winner justice when he's separated from the kids for a week?" asks Dennis Robertson, who manages convention services for Walt Disney World. Whether male or female, employees may be able to focus more completely on work when they know they have their family happy and at hand. "There is a greater public awareness in corporate America about the need to keep people happy at their jobs," says Glynna Hanchette, director of advertising and promotions for The Don CeSar in St. Petersburg Beach, Florida. "By letting the kids come, it's almost like an incentive. It shows that the corporate office is aware of how important family ties are. And, if you make the meeting more enjoyable, maybe people will learn more."

Hotels and resorts recognize the competitive edge they can gain by offering children's services as part of their large meeting package. Philip Wright, assistant general manager of the Buena Vista Palace in Florida, points out that "one of the biggest amenities a hotel can offer is not the soap or shampoo in the rooms, but rather the kids' programs. And it's an amenity that's going to be expected."

If you plan to include children in your next big meeting, you'll want to consider each of these suggestions from *Meeting News* based on interviews with Barbara Smith Nivala, National Speakers Association; Meg McLeroy, The Breakers; and Madeleine Dean, The Don CeSar:

Dos

- ▲ Decide whether to hire an outside firm, use an established hotel program, or run the program yourself.
- ▲ Plan well in advance. What kind of activities will be planned? How many children are coming? How old are they? What kind of schedule will you need (half- or full-day child care)?
- ▲ Decide who's picking up the tab. Make sure attendees know what costs they're responsible for.
- ▲ Obtain a signed release form from each child's parents. Consult with your legal counsel for content and wording of the release.
- ▲ Provide children with registration packets and badges. Their printed program should be included in the overall convention program so parent and child know where the other is at all times.
- ▲ Set an age limit on children. Remember that the younger the child, the more supervision is required, and therefore the more costly the program.
- ▲ Divide children attendees by age category: two-to-four, five-to-eight, nine-to-twelve, and thirteen and up. Give the younger ones structure; avoid heavily structured programs for teens. Do give teens a "hangout" place to meet other teens.

- ▲ Hire qualified, licensed supervisors. Avoid too many off-property activities to limit possibility of liability suits.
- ▲ Figure in time for mom and dad to be with the kids, even if it's just an afternoon. Otherwise you defeat the purpose of having children be part of your meeting.
- ▲ Do make the program educational. Consider tying in with established programs such as Walt Disney World's Wonders seminar series. Or, design your own program.
- ▲ Make the program fun. Combine recreation with education.
- ▲ Be careful about religious beliefs, particularly planning activities around holiday seasons.
- ▲ Discourage adults from too much involvement in the children's program. It creates more of a classroom setting, and kids are less apt to raise their hands and ask a question.
- ▲ Tie in with local attractions. For example, the Children's Hotel at the Peabody Orlando (Orlando, Florida) allow children to walk the famed Peabody ducks, playing honorary duck master of the day.

Don'ts

- ▲ Make children's programs mandatory.
- ▲ Serve gourmet meals. Opt for pizza, hamburgers, and french fries.
- ▲ Provide any entertainment that is not considered family-oriented.
- ▲ Think that holding a children's program limits your site selection. Even city hotels, including Washington's Hyatt Regency and New York's Plaza, have opened children's rooms. In 1990, the Denver Convention Center will become the first of its kind to open a day-care center for the children of convention attendees.

But you say you have no children? Or, you have children but look forward—with delight—to leaving them home for a few days of adult bliss at your meeting or convention? Then you may want to become active in helping your company plan for the unobtrusive presence of children at your next meeting. In the 1990s, count on the kids being there more and more.

2. LET'S MAKE THIS MEETING DIFFERENT

Network television makes much of the "sound bite"—that short, pithy burst of words that makes a good headline, echoes in the ear, and works well as a bumper sticker.

Managers, too, are looking for something with pizzazz for their meetings—something that will leap to mind when someone refers to "Frank Evans's sales meeting." They want to wake employees up to new possibilities, to shed some of the stuffiness that encrusts senior management, and to . . . well, have some fun. And no risk seems too great, no stunt too tasteless. Last year one senior vice president made his entrance onstage at a convention center by sliding down a wire suspended above the heads of attendees. The VP was decked out in a big diaper—"representing new things ahead for the company," he explained.

A West Coast real estate executive received from conventioneers a gag gift which he strapped on over his Brooks Brothers suit and wore throughout his welcoming speech—a pair of sequined angel's wings. In the conservative Midwest, an insurance senior executive clanked on stage wearing a full suit of medieval armor. Even the President of the United States doesn't mind getting down and different: Bush picked up an electric guitar at his inaugural ball and strummed along to backup provided by Lee Atwater, now chair of the Republican Party.

Done shrewdly, such moments can boost managerial careers and send team spirit zooming. Corporate leaders have trouble escaping the ivory tower in which they are often confined by the power structure, as interpreted in employee attitudes. When a leader voluntarily lets down his or her hair for a moment of self-directed humor or other hilarity, employees quickly come around to a new set of attitudes toward management: "She's a party animal!" "He's not such a stuffed shirt after all!"

Consider three ways that other companies have combined imaginative play with serious work at their meetings:

Meet on an Indian reservation. The Sunrise Resort, 200 miles north of Phoenix, lies in the middle of the Apache reservation, and is owned and run by tribe members. Charles Lutz, national sales manager, says the resort gets "a lot of business from government groups and executive getaways." Special features of this meeting site include not only proximity to the Petrified Forest and Indian ruins, but hayrides, Western cookouts, and performances by the famed Crown Apache Dancers.

Meet with the Muppets. Organizations as diverse as Citibank and the CIA have used two-to-four minute vignettes produced by Jim Henson Associates. Dozens of different video and film clips are now on the market starring familiar and not-so-familiar Muppet characters. Vignettes such as "Sell, Sell, Sell" and "Let's Have a Dam Break" make business points in a humorous, and usually satiric, way. The clips are an ideal way to begin, end, or break the boredom of a meeting. "They're such a complete switch that people listen better to the rest of the meeting," says Gordon Laird, marketing services manager at Exxon's Houston chemical division.

Divide and conquer: Meeting participants may often enjoy the camaraderie and gung-ho emotions that come from teams placed in competition. At a recent sales meeting, Prudential sent relay teams splashing across the Hawaiian surf in an innertube race. Seagram & Sons enacted a mini-Olympics, complete with the lighting of the symbolic torch. Libby-Owens Ford executives climbed on (and fell off) horses for a hilarious rodeo as part of a recent sales meeting in

Scottsdale, Arizona. "Salespeople are very driven," says Nancy Pavlik, president of Southwest Events, Inc., "and you want to give them something to channel their energy, but not something to make them look foolish."

The best kind of business play leads to a business point. At the fifth annual convention of the International Association of Conference Centers, attendees divided into four competing management teams. Each team had to write a manual, hold a press conference, and make a videotape "depicting the ideal sales presentation focused on the Conference Center Concept," according to directions. At the culmination of the exercise, the videos brought down the house—especially one team's Twilight Zone take-off, which twinkled on-screen as the Erogenous Zone.

There's rarely a problem getting even the most prestigious company members to participate. Pavlik recalls a goatmilking contest involving top executives from Libby-Owens Ford. "The most amazing thing," she says, "was watching these top business people on their hands and knees, stroking these goats to cajole them into producing more milk." The value of all this sloshbuckling is its ability to improve morale, open channels of communication, and increase team-work. "Employees who play together stay together," points out Georgetown University professor of management Dayle Smith. "The roles and norms that employees develop in social contexts influence the way they work together. Trust, support, and cooperation tend to be high following a successful social event."

You may feel that "it may work elsewhere, but not in my company." Don't be too sure. Start small by including just one mild moment of creativity in your meeting. Choose, perhaps, a few minutes from one of the business-oriented films and videos produced by John Cleese of Monty Python fame. Cleese's company, VideoArts, rents or sells more than 50 short and immensely entertaining videos for use in training sessions, conferences, and meetings. If your group appreciates your efforts to energize the meeting, include a discussion item on the agenda for a future meeting: "Suggestions for better meetings." Get consensus from the group on the need for something besides talk, talk, talk.

3. GEAR UP FOR INTERCULTURAL MEETINGS

You've worked next to José Ortega and Lisa Fong for three years. Because you all carry the same company ID, all park in the same lot, and all arrive and leave at about the same time, there's a tendency to forget completely José's Mexican-American heritage and Lisa's Japanese-American background. And what of the 18 Vietnamese employees in the company? Or the 47 Blacks, including 9 Muslims?

Although the individuals are invited, their cultural heritage is often ignored when it comes to large meetings. American business is especially parochial in missing the opportunity to celebrate the diversity of its workforce. And if we can't respond meaningfully in meetings to the cultural variety within our country, how can we prepare for an even greater challenge—meetings abroad?

In 1988, 34 percent of corporate meeting planners said they held at least one meeting at a venue outside the United States. Cynthia Wein, director of Creative Planning International, makes the point that "companies are becoming more aware that they are world citizens in a world economy." Attitudes and expectations on the part of meeting leaders must often be adjusted. In Italy, for example, "lateness" by American standards is common at meetings. Participants think nothing of getting up, walking around, and even leaving for short periods. Clearly, what plays in Peoria can't be expected to play in Florence, Osaka, and Amsterdam.

The key to successful intercultural and international meetings lies in taking opportunities, not sidestepping differences. Managers who plan and lead such meetings have a superb opportunity to

- compliment and involve cultural groups by including their food, music, and entertainment in meeting plans

- ▲ educate the whole company about its cultural groups through meeting events, print materials, and visual displays
- ▲ unite the company as one team by celebrating the variety of cultural groups comprising it

Admittedly, such cultural awareness can't be slopped on with a trowel. Active tokenism and paternalism are certainly as bad as passive insensitivity. To make a start in finding just the right intercultural activities and features to include in your next large meeting, make a cultural inventory of the company. What cultures are represented? Do these employees meet as a group inside or outside the company? Do they tend to seek one another's company at breaks and lunchtime? Do they take a relatively united stand on some company issues? Do they live in the same part of the city?

When you have assembled an accurate and insightful picture of your company's cultural diversity, think ahead to your next large meeting. In each of the following areas, how can you make room for and even highlight the cultural differences you've uncovered?

Greeting customs. How do your various cultural groups greet one another? Might such a greeting be appropriate or complimentary in a meeting?

Personal appearance and dress. Do your various cultural groups want to display their heritage and identity by particular ways of grooming or dressing? Are those styles permitted or encouraged for meetings? Would the meeting be richer if they were?

Religious attitudes. Will the entertainment, drinks, Sunday morning hike, or other supposed "attractions" at your meeting prove to be embarrassing to other cultural groups? Can you arrange "different strokes for different folks" in such a way that no one is offended?

Food. Have you chosen a menu that only White Anglo-Saxon Protestants can enjoy in good conscience? Could you pay a significant compliment to other cultural groups in the company by highlighting their foods on the menu? Can you arrange a variety of foods so that no one has to feel slighted?

Recreational activities. Have you inadvertently assumed that all company members, no matter what their culture, will enjoy the recreational activities you've planned? Will your seven women employees from India have fun passing the apple from neck to neck? Will your non-native speakers get the humor of "Rumor, Rumor," in which a message is whispered down a row? The question is not merely one of offense. Have you picked activities that other cultural groups can share?

Cultural slip-ups can scuttle business relationships. A group of 30 Tokyo business leaders visited the United States for the first time last year to meet with 30 American managers. Each American manager had been instructed to bring a gift for his Japanese counterpart—"Japanese always present a gift upon meeting a stranger," they were told. So each American manager dutifully arrived at the meeting hall with gift in hand.

Japanese managers, meanwhile, had been instructed by their cultural coaches to leave all gifts behind—"Americans do not present gifts when meeting a stranger." Though counter to their deepest impulses, the Japanese managers arrived at the meeting hall sans gifts.

You can imagine the immense discomfort that resulted. Americans felt the discomfort of the Japanese in not having a reciprocal gift, and the Japanese felt the chagrin of not obeying one of their oldest cultural habits in the face of American gift-giving.

For guidance on intercultural and international issues important to business, consult these sources:

The International Trade Administration
of the Commerce Department,
Office of Public Affairs, Room 4805,
14th St. and Constitution Ave., N.W.,
Washington, D.C. 20230

The U.S. Council for International Business
1212 Ave. of the Americas
New York, NY 10036

The International Meetings Guide, published by the Greater Washington Society of Association Executives. Available through the Washington Association Research Foundation, 1426 21st St. NW., Suite 200, Washington, D.C. 20036.

The Multinational Executive Travel Companion
Subscription Dept. REN-88
Harvard Square, Box 92
Cambridge, MA 02238

4. FLY THE FRIENDLY MEETING

Heading into the 1990s, major suppliers of services had discovered the "pipeline" approach to market expansion and diversification. Banks began using their mailing list to sell insurance; stock brokers were contacting clients about real estate investments. Sears pushed its Discover card and American Express marketed calculators and neckties in its monthly billings.

The airlines, too, were taking steps to provide more than a ticket stub and a lost bag. By the late 1980s, the best kept secret about meetings involved the airlines and what they could do to help you. If you're organizing a meeting for 10 or more attendees using air transportation, you'll waste money if you don't call the meeting services representative of your carrier (numbers below). Fares will probably be considerably lower than published fares, with fewer restrictions. But that's just the beginning of the story.

American Airlines, for example, keeps a Traveller Summary for meeting leaders, showing the arrival and departure dates and times for all meeting participants in addition to car rental information and hotel reservations. In its Meeting Place package, American offers its services in arranging for hotels, meeting rooms, food catering, AV equipment, and more. Eugene Rondeau, national manager for meetings for American, says his company "has taken the concept of service and subcontracted it to the very best suppliers. We have a deal for ground transportation, sightseeing, speakers, telemarketing, name badges, and housing management."

Other airlines also have excellent meeting programs, ranging from TWA's "10-Point Program toward More Manageable Meetings" to Delta's "Services for Company Meeting Planners." Most airlines will mail out a complete description of their services, with costs, and will answer questions by phone.

American Airlines Meeting Planning Desk, 1-800-AA-SALES
Continental Airlines Meeting and Convention, Sales Department, 1-800–527–3101
Northwest Airlines Meetings Desk, 1-800-328-2216
Piedmont Airlines Meetings Office, 1-800-334-8644
TWA Convention Desk, 1-314-344-3475
United Airlines Meetings Desk, 1-800-241-6522

Especially for larger gatherings, many companies have found that meeting packages put together by the airline are much less expensive than the do-it-yourself variety. It's worth investigating for your meeting plans in the 1990s.

ROLODEX RECAP

▲ many large meetings are making provisions for the whole family to attend, including children
▲ many companies are moving away from stuffy, formal meetings toward imaginative themes, activities, and locations

ROLODEX RECAP

▲ large meetings can compliment and involve a company's cultural subgroups by the selection of site, programs, food, and entertainment

▲ transportation companies can help you plan large meetings, often at advantageous rates

8

New Meeting Technologies

AMERICAN business spent more than $50 billion in 1988 flying bodies to meetings. In many cases, only minds needed to be there. Most of us know the stress of business travel—the bumpy flight to New York for a two-hour meeting, then back to the airport for a twice-delayed trip home. Why couldn't we simply talk to one another the way Peter Jennings talks to us—well-dressed, well-spoken, well-transmitted, and all the while a short ride from home.

That dream—"being there" without really being there—has fueled the off-again, on-again development of teleconferencing since the early 1960s. By the end of 1988, only a slight fraction—5 percent or less—of all American managers had appeared in an audiovisual teleconference. Telecommunication experts are not sanguine about that percentage increasing dramatically in the short term.

Why the fizzle? Ever since they were fascinated as children by Dick Tracy's video wristwatch, business people have wanted to speak face to face, not just mouth to ear. And technology has cooperated—teleconferencing rooms are available coast-to-coast in many hotels. Even the cost has begun to make sense: a 1-hour full-motion transmission L.A. to New York now costs less than $1,000. Try to fly a few executives to a cross-country meeting and back, with hotel bills, for that amount.

John Naisbett, for all the mega-hoots about his *Megatrends*, makes a telling point in his association of "hi-tech" with "hi-touch." No technology catches fire, Naisbett argues, without strong support from the human side—our preferences, needs, and habits. In the case of teleconferencing, business people haven't decided quite how to touch this dazzling new meeting tool.

To understand the general reticence to use teleconferencing for meetings, put yourself in the place of an executive outside the teleconferencing room. You know from previous tapings that you look lousy on video—ten pounds heavier and ten years older. And you're worried about feeling, sounding, and appearing unnatural before the camera. After all, meetings are supposed to "flow" through natural conversation, with interruptions, sounds of agreement or disagreement, and a thousand subtle signals of how things are going.

You have no feel for those subtleties as you look at another person's torso on the screen and try to talk to it. You feel wooden—memories of your first public speaking class come flooding back. You can't even shake hands with the guy.

And who else is in the room? Who's off to the side, out of view of the camera? Are you being taped? Is someone along the way—a satellite dish hacker, maybe—intercepting this supposedly private business conference?

All of these problems—personal discomfort, audience ambiguity, and a perceived lack of security—keep most business people from trusting important business negotiations and meetings to teleconferencing.

Avon's 100th anniversary meeting in 1986 makes an interesting test case. Avon faced the daunting task of assembling 2,000 district managers to discuss strategies for selling during their fourth quarter. Instead of flying these managers to one location, however, Avon opted to link five locations by interactive video teleconference. Their expectations of the video experience were high. "We held a videoconference for the first time," says Avon meeting manager Linda Livingston-Mugnos, "because we wanted every member of the Avon sales management family to be together simultaneously and share the excitement coming out of the celebration."

Some aspects of the excitement, however, just didn't translate well on the airwaves. The fashion-show segment, for instance, was live in New York—and simultaneously live also in the four other locations. Models in each city wore the same number of outfits and walked down similar runways at about the same moment in time. In other words, you had to "be there" for this event.

When it came time to publicize the gala event to the next echelon, Avon's 61,000-member President's Club, the mission wasn't entrusted to teleconferencing. Jim Preston, Avon's president, personally hit the road to press palms, slap backs, and show his meeting videotape to 33 cities in 21 days.

The point is simply this: teleconferencing, for all its apparent usefulness as a meeting technology, just hasn't been embraced whole-heartedly (or even half-heartedly) by American business. That acceptance awaits a re-education of managers in skills they didn't learn in B-school:

- how to sit, stand, and look before the camera
- how to use speech patterns, pace, volume, and emphasis that work well on video—and how to suppress personal mannerisms that don't
- how to participate naturally in the give-and-take of televised conversation, without awkward pauses, voice-overs, and mis-starts
- how to use time to advantage, creating "speech bites" in a way never required in ordinary conversation

These are the "hi-touch" aspects of teleconferencing that can't be solved with new chips or bigger satellite dishes. Human resource directors in major corporations are beginning to include executive training in teleconferencing as a high priority for the 1990s. If your company hasn't yet heard the gospel, a good place to begin is Lillian Brown's *Your Public Best* (Harper and Row, 1989). Ms. Brown has taught CEOs, legislators, and even presidents to put their best foot forward electronically. Her advice is worth practicing.

COMPUTERS AND MEETINGS

"I go to meetings to get *away* from computers!" an accountant told me a few weeks ago.

But there they are—two or three on the conference table. Michael Schrage, writing in *The Wall Street Journal,* calls these interlopers "the most useful tool for running meetings since Robert's Rules of Order." Computers, he says, can function as "intelligent chalkboards." The process works like this. A quick typist (not necessarily a secretary) is given the chore of typing up key points made by participants as they talk. These points appear on screens around the room for all to see.

When participants "see" the flow of their ideas, connections and associations begin to form:

"Hey, this point relates to something that was said earlier," or "Wait a second. Doesn't that contradict what you said ten minutes ago? Let's see. . . ." With a tap of a few keys, the screen scrolls back to the desired spot on the list of ideas.

Using commercially available software, meeting participants can see their comments in the form of an outline or in alphabetic/numeric order.

Of course, the computer can also display more complicated information in the form of spreadsheets or graphics. Unlike data on flip charts, the information displayed by the computer can be changed within a few seconds, with an adjusted spreadsheet or graphic at hand almost immediately. All information appearing on the screen can easily be printed out if necessary for "hard copy" distribution to the group.

Schrage sees a fascinating sociological aspect to this use of computers: "Participants meeting in this environment often pay more attention to the points displayed on the screen than to the individual making the point. What's on the screen is treated as a creation of the group—not personal property. This technique apparently filters out some of the 'dueling egos' mentality that infests too many meetings."

Two recent software products—*Lotus Notes* and Broderbund's *For Comment*—allow meeting participants (whether physically present or not) to shape a common agenda, document, or other joint effort. Participants guide the way the communication grows by offering comments in marginal areas of the screen. Those suggestions can be accepted by the group and incorporated into the growing communication, or rejected.

Finally, computers are a godsend for scheduling meetings. Using electronic mail or a computer bulletin board, various potential attendees can "check in" with available time slots on their calendars. The computer sorts out the best possible meeting times and informs participants accordingly. "Besides," says a Texas insurance underwriter, "it takes the favoritism and politics out of meeting times. If the computer says we meet at 10 A.M. on Tuesday, none of us can claim the boss is oblivious to our individual schedules."

REPRODUCING GREASEBOARDS

Several Japanese companies and one American firm now produce greaseboards that pump out photocopies of anything written on them. Depending on screen size and photocopy capability, these marvels cost from $1,600 to $5,000. Some models feature screen "memories" that allow the board to be erased and written over several times, with each separate screen recoverable at the end of the session by photocopies.

PRESERVING IMPORTANT MEETINGS

It's ironic that we preserve photocopies of even the most insignificant memos— but rarely preserve actual "footage" of expensive company meetings, often in exotic locations. Howdy B. Letzring intends to change that. His company, Wild World of Sports, filmed nearly 50 corporate meetings in 1987 for such companies as AT&T, Philip Morris, and Mastercard International. The company's name stems from a common theme in its video productions—corporate tennis or golf tournaments. Letzring and his crew will produce a video combining meeting highlights with outstanding (and often hilarious) moments from company sporting events.

Corporations have used such videos for in-house motivational meetings, orientation for new employees, and marketing purposes.

TURNING SOUNDS INTO PRINT

You've chosen one of your company's finest to prepare a presentation for an upcoming meeting. You've provided resources—a lot of resources, in fact, when salaried time, support staff, and research funds are tallied. The moment comes, the presentation is a smash success, and then . . .

And then nothing. Too often, well-researched and well-spoken words disappear into thin air after being delivered at meetings. Jean Horiszny, director of Technical & Trade Press Relations, asks, "Why not get a 'second crop' out of that investment by reworking the information into an article for an appropriate trade publication?"

Horiszny calls meeting presentations "a natural source of articles" for four reasons:

- ▲ A meeting paper generally presents something new, often a good basis for a story.
- ▲ A meeting paper is generally targeted to a particular audience, the same as a trade publication.
- ▲ The research is complete. . . . Approval has already been given for its announcement, and for the precise nature of the information and phrasing.
- ▲ Illustrations have often been prepared.

The company stands to influence a much wider audience—clients, potential clients, present employees, and the general public—by putting presentations into print. Usually such publications are a source of pride both for the author and the company.

JUST AROUND THE CORNER

The 1990s will see the commercial use of a revolutionary computer technology already in prototype: the voice-driven computer. It may be possible by the end of this century for all words spoken in a meeting to be entered silently by the computer into a data base for storage and analysis. As companies come to depend more and more on artificial intelligence and expert systems (what some have called CAT—Computer Assisted Thinking), human meetings will become important input occasions by which powerful computers are "taught." A business presentation in that day will be given to a human primary audience and, by voice-driven computer technology, to computers as a secondary audience. No doubt we will learn to choose our words accordingly.

Ironically, the same computers that make mass communication possible may also help us make communication more individualized and personalized. Alvin Toffler, author of *Future Shock,* foresees "smaller, more intimate meetings" in the future. "There comes a point," he says, "at which the delivery of information is less effective when it is more 'massified.'" As an example, he cites his own joy and candor when speaking to a group of ten, and the feeling of discomfort and separation when speaking to thousands.

The computer can aid in the coming individualization of meetings by keeping track of who-said-what-to-whom, when, where, and why—and what it all means to the ongoing progress of the project or enterprise. Without computers, big meetings were necessary to keep everyone informed. With computers, small meetings are possible while still keeping everyone informed.

Toffler's example of such an electronic beehive of separate cells is Digital Electronics Corporation: "At DEC you have very large numbers of engineers and employees on electronic networks constantly communicating with each other around the world. There was the case where there were something on the order of 300 engineers around the world working jointly on some project, and they decided among themselves they needed face-to-face communication. Then, without going up the hierarchy for approval, they created a meeting and said, 'Let's all get together in Geneva next month.' . . . At DEC they've discovered—and I think it's perfectly understandable—that electronic communication needs to be reinforced by face-to-face communication. . . ."

ROLODEX RECAP

▲ for effective audiovisual teleconferencing, participants must be trained to use this new meeting technology
▲ computers can be used at meetings as an electronic chalkboard to display developing ideas
▲ reproducing greaseboards can produce photocopies of material written on the board

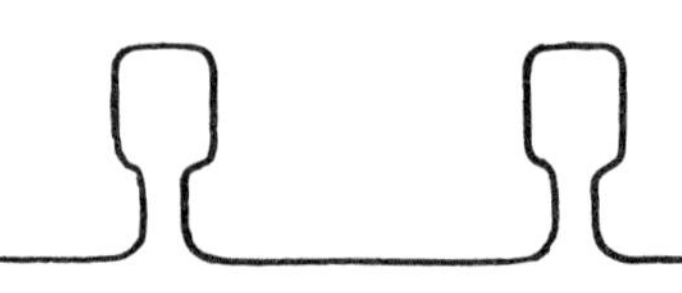

ROLODEX RECAP

▲ important meetings can be preserved on film, often in an entertaining way by blending a company sports event with company business
▲ especially well-conceived meeting presentations can be converted into articles for trade magazines and journals

ROLODEX RECAP

▲ as new technologies emerge, shrewd managers will look for appropriate ways to apply them to more effective meetings

9

Better Training Meetings

WHY single out training meetings for special attention? First, because they are the fastest growing form of meeting in American business. By the year 2000, it's estimated that between 5 and 15 million manufacturing jobs will be restructured, with another 16 million new jobs being created. The National Alliance of Business forecasts that we will experience as much workforce displacement in the next twenty years as we have, as a nation, in the last eighty years. As the companies reposition themselves to take advantage of new technologies and new markets, millions of workers must be trained or retrained.

Second, training meetings need special attention because they are, in many companies, just plain lousy. The symptoms are everywhere: training resources are severely restricted, training materials are poor, trainers themselves are lackluster employees, and trainees approach the experience with dread. Roderick Wilkinson, writing in *Supervision*, Oct. 1988, points to the difficulty of finding a "training leader . . . who knows when to stop lecturing. . . ." The unfortunate fact, says Wilkinson, is that "nobody pays much attention to a lecturer after 15 minutes unless he [or she] is very, very good. . . ."

THE USUAL TRAINING MEETING

We've all been there too many times—the training meeting. It's held in a cramped, musty room reminiscent of Ms.

Grundy's fourth grade. Desks or tables are arranged in straight rows and columns, to set the stage for the kind of thinking that will go on in the room. Learning materials are bad photocopies from a bad mimeograph prepared on an ancient typewriter. (We know the trainer's first, inevitable words: "I'm sorry these pages are a little hard to read.")

Ah, the trainer—hardly a major player or significant talent in the company. He or she probably liked to "play school" as a youngster and has never given it up. The voice drones on and on as, on cue, our pages seem to flip almost by themselves. We stare blankly at the words and diagrams, then at the trainer, who stares at the ceiling or back wall. Minutes ooze slowly by. We're being trained, all right—trained to think of our jobs, our company, and ourselves as terribly boring. Two more hours to go.

Sound familiar? It doesn't have to be this way. Getting up to speed for new tasks and responsibilities should be one of the most energizing and challenging occasions in business. Trainers should be the liveliest, most gung-ho spirits in the company, not its dullards.

SETTING GOALS FOR THE TRAINING MEETING

After all the pedagogical and psychological jargon is stripped away, too many trainers believe their mission is simply "to tell others what to do." Everything in a bad training meeting flows from this narrow point of view. The trainer, as "teller," stands up front. The workers, as "listeners," are seated like an audience so that all look forward to the teller. Information flows one-way except for infrequent questions.

In fifteen years of consulting for more than thirty Fortune 500 companies, I've come to know many superb trainers who set much more sophisticated goals for themselves and their students. Here's a loose summary and paraphrase of some of those goals:

> "I try to help trainees grasp how to learn, not just what to learn. After the training session, they'll be on their own. I

won't be there to spoonfeed them. They'd better know how to be their own teacher."

"I want each of my trainees to know why their training is important. Once they understand how much their success matters to the company and to their own futures, they really give maximum effort."

"I see a training session as an excellent time to network—to get to know other team members in the company who have special abilities, contacts, and resources. After the training sessions, it's those people you'll turn to for help in getting the job done."

"In my training sessions I do everything I can to build morale and esprit de corps. Our exit interview data show that the company's biggest problem isn't training people, but holding on to people we've trained. So a major purpose of our training program is to make employees feel they are part of the corporate culture—that they matter to the company as individuals."

When these assumptions are thrown into the mix, the training meeting looks, sounds, and feels much different from a lecture. The doors are thrown open for a wide variety of stimulating methods of communication and learning. Consider trying some of these new approaches in your next training meeting.

INNOVATIVE TRAINING TECHNIQUES

1. Programmed Training

Good training leaders often burn out after saying approximately the same words for the umpteenth training meeting on the same topic. "I heard myself explaining what a sentence fragment is one too many time," says a former business writing trainer in California.

Programmed training provides a way to clone the trainer's approach and to adapt it to individual learning styles. The content of the training meeting is reduced to "frames" that present information and ask for trainee response. Here are a few typical programmed training "frames".

These frames can be delivered to trainees by booklet, audiovisual slides or transparencies, and computer. In all these cases, the programmed answer to the questions is revealed

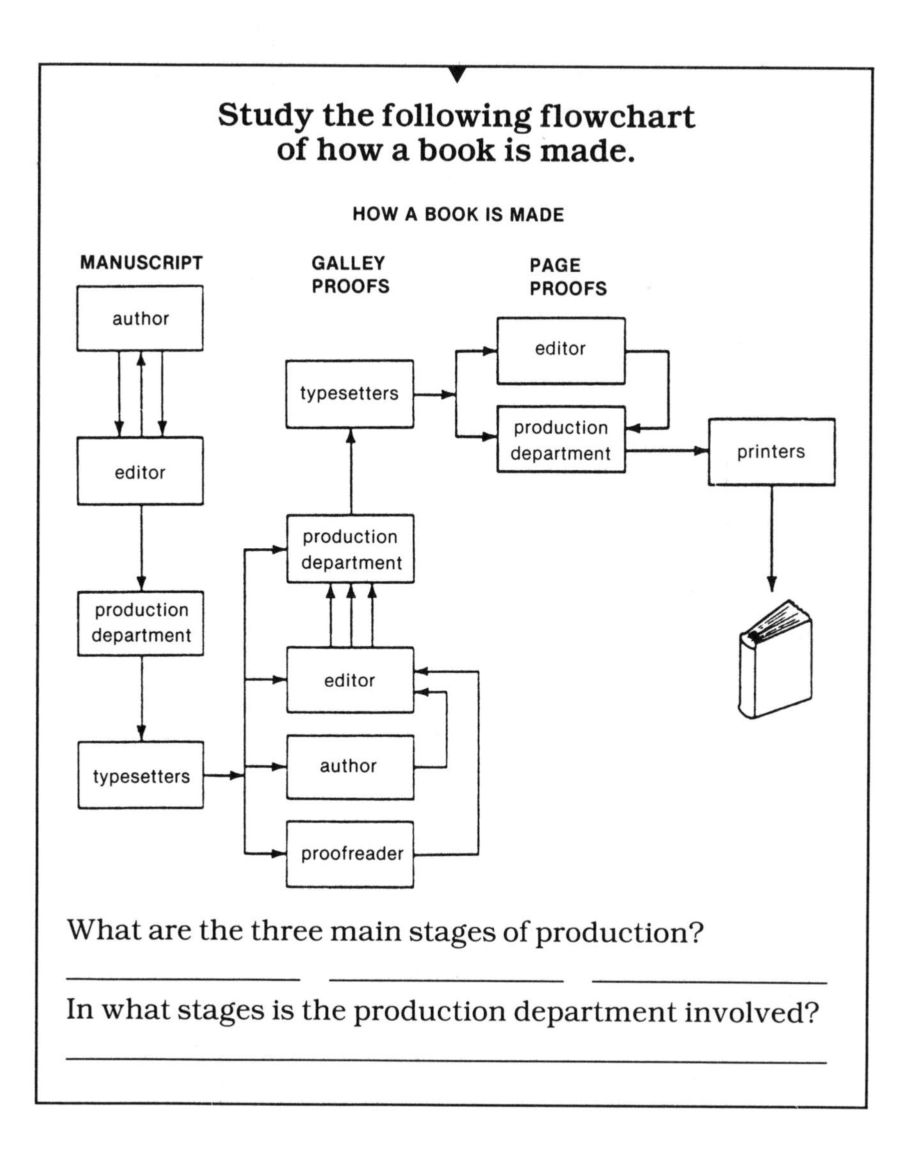

immediately after the trainee has entered his or her response. In a branching program, the trainee sees remedial frames to "recover" from a wrong answer. In a linear program, the trainee simply forges ahead, noting where his or her answers differ from the programmed answers. Most training and development departments require, in trial tests, that about 85 percent of the frames be answered correctly for minimally acceptable levels of programmed training.

Few training meetings, of course, are given over entirely to programmed learning. But it may come as a welcome change of pace in a meeting dominated by verbal exchange to pass out a short programmed exercise: "Instead of telling you about the new project, let's work through a short programmed exercise that will help each of you locate what you already know and still need to know."

2. "Hit List" Discussion

In this form of training, participants are asked to come up with an agenda or "hit list" of topics they want to discuss. Often, the topics selected won't be identical to the trainer's list—but so much the better. Trainees should have some way of letting the trainer know what "took" and what's still confusing.

The "hit list" activity can be announced as part of a meeting spent primarily on new information: "I'm going to give you a quick overview of the project. Then I'll take notes on the board as you tell me what areas you'd like to discuss." It's important to formalize the "hit list" in writing so that all concerns on it are addressed, and to prevent the trainer from veering off to his or her favorite topics.

3. Panel Discussions

Three or four voices are infinitely more interesting than one. A panel discussion brings several points of view to bear on a training topic. Panel members usually begin with a short statement encapsulating each person's approach or perspective. Then the panel discusses or debates the implications of their similarities and differences. Audience members can ask questions or make comments at any time.

At a large Michigan carpet distributor early in 1989, a panel discussion was used to orient new employees to hierarchies of management. The panel was made up of a company vice president, the director of accounting, the operations chief, and the sales manager. Each talked about his or her individual responsibilities, then all talked about administrative challenges facing the company. New employees got to see firsthand—not just hear about—who did what in the running of the company. They got to know the faces and personalities behind the names they had seen on memos.

4. The Trainee Panel

A variation on the panel approach uses trainees themselves as panel members. Given individual assignments and time to prepare, trainees enjoy the chance to "play expert" before the group. The group, who has also prepared the assignment, keeps the panel honest by asking questions and pointing out mistakes.

This training technique can be introduced naturally as part of a training meeting: "For the next 20 minutes or so I've asked six members of the group to give us their perspectives on the topic. After their presentation, we can ask them questions."

5. Modeling

This training technique emphasizes showing, not telling. McDonald's uses modelling for its trainees by showing film clips and live demonstrations of "ideal" employee actions. A food server is seen, for example, holding the "to-go" package in two hands, not one, while thanking the customer. Trainees are then given a chance to imitate the model.

The "whys" for desired patterns of behavior are often not explained: "This is how we do it here." Or, discussion of the modelled behaviors can follow the trainee's attempt at imitation. A company might use modelling to train new interviewers to practice certain verbal and nonverbal techniques in interviews. Only with growing mastery of the techniques does the trainee learn why they are important for the company.

6. Case Studies

Spurred in part by the immense popularity of the "Harvard Case Approach" to business topics, case studies are appearing more and more often in training meetings. Some case studies are "closed"—they tell a completed business story, and ask participants to discuss such issues as the merits of the decisions made by characters in the case. Other cases are "open," presenting a business scenario or dilemma that must be resolved by participants, often divided into teams.

Case studies give trainees a chance to apply learning in a "safe" environment. Their decisions, right or wrong, won't cost people their jobs or hurt the company—it's only an exercise. But the analytic processes and group dynamics involved are very real. Those same factors will determine success for trainees in later, actual decisions.

ON THE JOB TRAINING

One particularly interesting way to present a case is to show the opening segments of one of the films or videos listed at the end of this chapter. Turn off the presentation just at the point that the dilemma reaches the point of crisis. Then let trainees invent a resolution. After discussing their insights, you can turn the projector or VCR back on to see how Hollywood handled the problem.

7. Fortune Cookies

This variation on the case approach makes use of a series of brief messages—hence the title—to guide discussion and decision-making. It works like this: All trainees read, hear, or see the same introduction to a business situation. After this common experience, trainees are divided into several teams and given a multiple-choice question such as the following:

> You have watched Frank ask six employees what they know about a recent theft of company equipment. Now Frank should . . .
>
> A. Drop the matter because of insufficient evidence.
> B. Communicate his suspicions about certain employees to his superiors.
> C. Turn the whole matter over to company security without comment.
> D. Call in two or three of the most likely suspects for a more intense questioning.
> E. Call a general staff meeting to put peer pressure on the guilty parties and to solicit new leads.

Each team is asked to settle upon the choice they like best (even though the team may favor another, unnamed choice). Based on their choice, the team then receives the second "fortune cookie"—a brief extension of the situation. If the team chose "C," for example, the extension message might be

> Security acts promptly by calling each of Frank's employees in for a grilling. Morale plunges in his unit as employees grumble about "being accused" while Frank, the boss does nothing

to "stand up" for them. Faced with a degenerating work situation, Frank should. . .

(five more choices follow)

You can arrange your choices and consequences in any way you wish. The value for teams lies in the discussion and decision-making process. Managers involved in such activities often learn to know one another's instincts—who's a hard-liner, who's a good negotiator, who's a "people person," and so forth.

8. In Basket /Out Basket

In this meeting exercise, each participant receives a folder of identical "in basket" items—for example, a report showing abuse of telephone privileges by employees, a list of employees absent or late-to-work during the last month, an invitation to speak at a charity event, a memo with two questions from a superior, etc.

Each participant is given the same amount of time—45 minutes, perhaps—to prioritize these items and decide how to handle each of them. If typewriters or word-processors are available, participants can be asked to generate responses for the top three or four items in their prioritized list. Valuable discussion can follow, focusing on such issues as how priorities were determined, what responses were chosen, and why.

9. Role-Plays

Nothing is more boring than listening to a trainer describe human traits, such as the likely responses of a customer to a sales presentation. How much more interesting and memorable to see those responses "live." Using role-plays, one or more experienced company employees can take on roles they probably have come to know well over the years—client, supplier, distributor, and so forth. Less experienced participants (the trainees) can interact as company representatives with these players.

Role-plays can be center-stage, with all eyes on a single encounter at the front of the room, or distributed in simultaneous encounters throughout the room. When it comes time for discussion, characters can be quizzed in their roles: "Mr. Client, how did you feel when our representative started pressuring you for a yes/no decision?" These sorts of in-role queries can often yield more candid and more interesting responses than asking participants to philosophize on sales techniques.

There's often some resistance to the "drama" of role-plays in meeting environments where they've never been tried. Therefore, plan your first role-play with care. Choose sure-fire participants who can demonstrate how the technique gets to the heart of issues in a lively, interesting way.

10. Fishbowls

In advance of the meeting, draw up a list of eight or ten important questions or concerns about your chosen topic. Divide meeting participants into the "ins" and the "outs." The ins take their places inside the fishbowl, with the outs forming a circle around them:

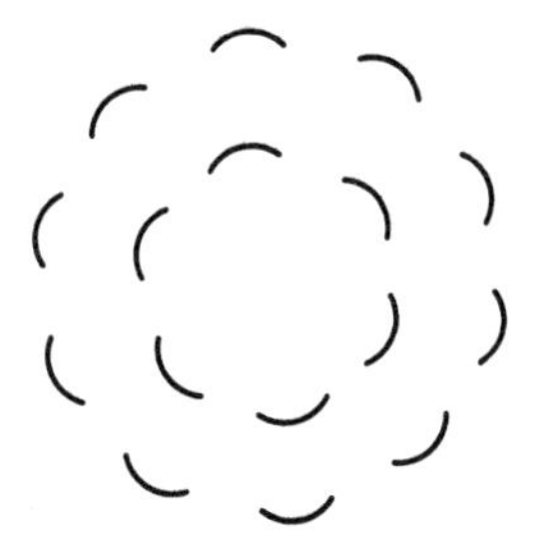

The ins are given the list of questions with the instructions to discuss any or all of them in the time allotted (perhaps 20–30 minutes). During this period, the outs listen carefully, taking note of key points, arguments, examples, and details. When the time period has expired, the outs speak up for 5 minutes or so to give the ins feedback: what was said, what wasn't said, what was emphasized, who did most of the talking, and so forth.

Roles are then reversed for 20–30 minutes, with the outs taking the inner circle and the ins the surrounding chairs. A final debrief of all participants concludes the activity.

Using the fishbowl as a meeting technique forces some of the biggest mouths in the room to become ears for a change. It also encourages more timid and introspective types to speak out. The feedback process gives participants a chance to correct misperceptions: "what you heard is not what I meant to say."

SUITING THE TECHNIQUE TO YOUR MEETING OBJECTIVE

Such creative approaches to training meetings shouldn't be chosen simply for variety (although even that reason would be welcomed by many trainees!). Ideally, the technique you choose should grow directly and naturally from your meeting objective.

As in the following example, objectives for training meetings can be stated in general and behavioral terms. The general objective proposes goals that are often long-term and difficult to measure. Behavioral objectives, by contrast, state what trainees will be able to do (in observable ways) after training:

General Objective

- Trainees will feel a sense of partnership and team spirit with other company employees in meeting production goals.

Behavioral Objectives

- Trainees will use the digital access system correctly to gain entry to master computer files.
- Trainees will name the eight product lines of the company correctly, with the three major characteristics of each product line.

In stating behavioral objectives, the trainer should make trainees aware of the criteria used to measure performance.

In the case of the digital access system, performance will be measured by the number of incorrect entry attempts logged by the computer from trainees. The success of training, that is, can be measured by a known standard. In the case of the eight product lines, employees may be asked to fill in the blanks on a written test. The trainer will measure success against the criterion of the company's list of products and their characteristics.

SHOW THE TARGET

Both general and behavioral objectives should be stated in commonsense terms at the beginning of the meeting (or before, in a meeting announcement). Participants want to know what they'll get out of the meeting (the general objectives) and what they'll be able to do afterwards (the behavioral objectives). Here's how one manager cut through the gobbledygook to tell meeting participants what to expect:

> "This afternoon we're going to give you a feel for how decisions are made within work groups in the company [*general objective*]. After participating in buzz groups and a brief panel discussion, you will be able to identify which manager to see about a range of common problems; how to communicate directly with upper management; and how to check on what's happening with regard to your questions or concerns [*all behavioral objectives*]."

YOUR OWN REPORT CARD AS MEETING LEADER

How well you trained participants for behavioral goals will be measured empirically by computers, tests, quality control inspections, and time clocks. But to what extent you achieved your more general goals may be more difficult to monitor and measure. How can you assess participants' attitudes about the training meeting? Ask them.

Here is a useful evaluation form that will tell you at least three things about your training meeting: what you did well,

greatest reliability of responses, the anonymity of respondents should be stated explicitly.

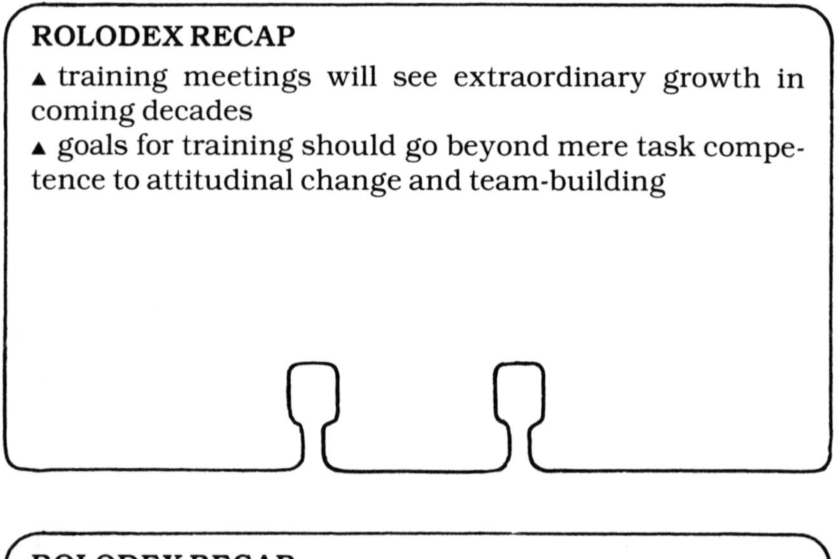

ROLODEX RECAP

▲ training can be made more individualized and often more lively by programmed instruction, "hit list" discussions, panels, modeling, case studies, "fortune cookies," in/out basket activities, role-plays, and fishbowls

Evaluating a Training Meeting

1. In your own words, state the purpose of the training meeting you have just attended. ________________________

____

____

2. How well do you think the meeting achieved its purpose? ________________________________

____

____

3. What could have been done to improve the training experience for you? ________________________

____

____

   For others? ________________________________

____

____

4. How effective was the use of time during the meeting?

____

____

   What topics or activities deserved more time? __________

____

____

   What topics or activities deserved less time? __________

____

____

5. Please evaluate your meeting leader or trainer. ________

____

____

   What suggestions can you offer to help your meeting leader or trainer meet the needs of the group? __________

____

____

6. Evaluate your opportunities to practice new skills or apply training learned in this meeting. ______________

____

____

7. Were you evaluated fairly and accurately? __________

____

____

   What suggestions can you offer to improve the evaluation process? ______________________________

____

____

8. Evaluate the training site. ______________________

____

____

9. Evaluate the usefulness of the training materials. _____

____

____

10. Evaluate the group chosen for this training meeting. _

____

____

   Was the size of the group appropriate? ___________

____

____

   Was the composition of the group appropriate? _______

____

____

Appendix: A Case in Point

Lisa Chen supervises fourteen technical writers and graphics/production specialists in the high pressure/high production Manuals Unit of a major computer manufacturer.

Over the years, Lisa has seen to it that communication links in the unit are superb. Employees are connected to Lisa and to one another by electronic mail (via computer terminals), voice mail (via telephone), computer bulletin boards, and—for employees separated from the group by travel—teleconferencing capability. "Good communication channels make it possible to avoid a lot of meetings," says Lisa. "I can usually give and receive the input I need from individual employees or groups by using the computer or phone."

But today Lisa seriously contemplates calling all 14 of her people off their assigned work for a 90-minute meeting. She calculates that this gathering will cost the company well over $2000 in salaried time and resolves to make the meeting worth that investment.

REASONS FOR A MEETING

The meeting is necessary for four reasons, she believes. First, senior management has fallen in love with a competitor's manuals—bright, cartoon-filled booklets that supposedly "make learning fun." Lisa wants all her employees to hear of this new interest on the part of senior management directly from her, not from the grapevine. It's a sensitive topic, one that she doesn't want to commit to a written memo. She wants to be there to make sure her employees understand that senior management isn't unappreciative of their work on previous manuals.

Second, she needs input from her people on what they consider the merits and demerits of the "cartoon" approach to manuals. Senior management will soon ask her, as manager, for her opinion on the matter. She wants to be as informed as possible, and to speak with the backing of her employees.

Third, the unit has three new employees. Lisa wants to introduce them in the meeting, describe their areas of responsibility, and make them feel part of the group. She wouldn't call a meeting just for this purpose, but it's one more reason to get together.

Finally, Lisa wants to meet to reinforce her role as leader. Electronic mail messages are fine, she feels, but they do little to give employees an idea of her leadership style. She wants them to know her as more than a signature.

GETTING READY TO MEET

Lisa wants her people to do their homework prior to the meeting. A week in advance of the meeting date, she sends out the following meeting announcement memo:

To: Manuals Unit employees
From: Lisa Chen, Manager
Subject: Preparation for June 11 meeting
Date: June 3, 199_

Please mark your calendars for a 90-minute meeting beginning at 10 A.M., June 11. Our main business will be discussing new ideas for manual design for the coming year.

In preparation for that discussion, let me ask each of you to review competitor's manuals (most are on file in our Manual Library) and to gather your own ideas for improving our manual design and production.

I'm in the process of drawing up the meeting agenda. If you have items you want included, please contact me no later than close of business on June 6.

Encouraged by this memo, several employees do offer agenda items. Some are appropriate for the meeting and some are not—these latter items Lisa tries to take care of by a phone call or delegation. One employee's suggestion is especially useful for Lisa's purposes. One of her technical writers suggested that "we discuss how to make our manuals more lively—maybe with more drawings or even cartoons." Lisa calls the employee, thanks him for the agenda item, and asks him to prepare a short introduction to the idea at the meeting. In this way, Lisa avoids the role of "heavy" in telling her employees what senior management wants. If the idea for change can come from within the group itself, so much the better!

THE AGENDA

One June 7, Lisa distributes the following agenda to her employees. Notice that agenda items are described in some detail, with specific people named to "carry the ball" of presentation and discussion. Lisa will play the role of facilitator rather than tyrant in the meeting. Suggested time limits have been placed beside agenda items to make sure all items are discussed, with appropriate prioritization of time.

Agenda

Meeting: 10 a.m., June 11, 199_
Place: Conference Room A

Attending: All Manuals Unit employees
Information contact: Lisa Chen, ext. 2312

Purpose of the meeting: to discuss the development of new approaches to manual design and production.

Suggested preparation: review competitors' manuals, gather your best ideas for improvements

	Item	Suggested Time	Input Person	Action Taken	Action Person
1.	Introduction of new employees	10 min	Marg. V.		
2.	Review of competing manuals	30 min.	Mike M. Fred C. Teresa W.		
3.	Analysis of "cartoon" approach	10 min.	Lisa C.	*(this portion filled in during meeting)*	
4.	Alternate design approaches	20 min.	Gwen Y. Itak P.		
5.	Formation of an ad hoc committee for discussions with senior management	10 min.	Lisa C.		

THE MEETING

Thanks to preparation—Lisa's and the participants'—the meeting goes well. Everyone has a chance to speak up, and divergent points of view are aired in a constructive spirit. The group reaches consensus on the necessity to move toward more colorful, motivating manuals. They decide to stop short, however, of imitating the competition's "comic book" approach, feeling that users want clear, organized information more than entertainment in manuals. Lisa promises to convey the consensus opinion of the group to senior management.

FOLLOW-UP

After the meeting, Lisa takes time to write a brief thank-you memo to meeting participants. She wouldn't do so after every meeting—the practice might cloy—but she feels such a memo is in order, given the obvious preparation and hard work of the participants.

To: Manuals Unit employees
From: Lisa Chen, Manager
Subject: Thanks!
Date: June 12, 199_

As many of you know, meetings are not my favorite activity in this company. A good meeting is especially gratifying—and I thought we had a particularly good meeting on June 11. We did our homework, aired differences, stuck to the topic, and reached consensus. We even adjourned on time! Thanks for your help in making this time together worthwhile. Minutes of the meeting will be distributed on June 14. Please read them carefully to check on assigned follow-up tasks.

THE MINUTES

Lisa gave the task of taking the minutes of the meeting to Bob, her secretary. She didn't want any of the unit's other employees to give less than full attention to discussion. Besides, Bob is trained in taking quick, accurate notes. She proofreads Bob's draft of the minutes, then distributes it to her employees. She also posts it on a computer bulletin board and on a wall bulletin board to let other work units know what's happening. Two senior managers (those most interested in the competitor's cartoon manuals) receive confidential copies of the minutes with an accompanying note from Lisa—"We're acting on your suggestion regarding brighter, more motivating manuals. Let's get together at your convenience to discuss several exciting ideas from the Manuals Unit."

The minutes are written in a natural style—they are, after all, intended to be read!

Minutes

Meeting time: 10 a. m., June 11, 199_
Place: Conference Room A
Agenda: (attached)
Attendees: All Manuals Unit employees in attendance

Lisa Chen welcomed meeting participants. Her opening remarks focused on the expanding responsibilities of the Manuals Unit in supporting company products. With such expansion has come the need for three additional employees—Jim Franklin, Bess Ortega, and Norman Lee. Each was introduced in turn by Margaret Victors.

Discussion turned to the main topic of the meeting—ideas for more lively, colorful, and motivating manuals. Mike Morgan, Fred Conway, and Teresa Wilson summed up their research into competitors' manuals, particularly those with "fun" formats. Samples of such manuals were distributed by Lisa Chen. The group felt the "cartoon" approach to computer manuals had strengths and weaknesses:

Strengths:

- appeals to young computer user
- gives "user friendly" appearance to product
- motivates use of manual (instead of calls to technical support)

Weaknesses:

- makes product appear to be a toy
- may alienate serious business user
- makes organization and reference system within manual difficult
- doesn't harmonize well with general company image
- has been used extensively by competitors—our efforts may be seen as mere imitation
- may not be necessary, as computers become more and more familiar to general public

It was the consensus of the group that new manual development should focus on motivational design components used by upscale national magazines instead of the "comic book" approach of competitors. Gwen Young and Itak Pae summarized the high-interest design components of national magazines as:

- ▲ prominent use of interesting faces
- ▲ action photographs
- ▲ artistic use of unusual colors
- ▲ development of graphic themes carried out throughout document
- ▲ use of modern fonts and typographical elements
- ▲ reduced numbers of words per page
- ▲ substantial use of headlines, headings, subheads

Bob Walker and Sam Brown were asked to prepare a written report on such design elements to be sent to each employee no later than June 20. An ad hoc committee on new manual design (Barbara Quinley, Henry Port, and Everett Riley) agreed to assist Lisa Chen in presenting the Manuals Unit's ideas to senior management, if asked to do so.

The meeting adjourned at 11:30.

Respectfully submitted,
Robert Preston

Lisa feels her meeting has accomplished both overt and covert goals. On an overt level, the Manuals Unit has learned that new approaches to manual design are needed. On a covert level, the group has been motivated to develop those new approaches from within rather than having them imposed from senior management. Lisa feels that the group will be more productive if they have some degree of "ownership" of these new ideas. Also on a covert level, she has physically reinforced her position as group leader. While she didn't dominate discussion, she did demonstrate her ability to guide debate, listen intensely, and respond meaningfully. The group's confidence level in her leadership has increased. Finally, the group got to see itself as just that—a group, a corps, rather than a number of individuals separated by office partitions. This sense of belonging would be especially important to the new employees in the group.

Because her meeting was prepared for, productive, and relatively short, Lisa knows that the group will look forward to the next meeting. When will that next meeting take place? She responds, "When it's needed."

References

Abik, Clem. CANADIAN BUSINESS, Nov. 1986, p. 77.
Augur, B.Y. SUPERVISORY MANAGEMENT, Aug. 1980, p. 36.
Bailey, Julie. WORKING WOMAN, Aug. 1987, p. 68.
Barmash, Isadore. NEWSWEEK, Sept. 9, 1985, p. 12.
Baron, Eric. BUSINESS MARKETING, Aug. 1987, p. 42.
Barron, Robert A. UNDERSTANDING HUMAN RELATIONS. New York: Allyn and Bacon, 1985.
Beldock, Donald. FORBES, Nov. 2, 1987, p. 208.
Bing, Stanley. ESQUIRE, April 1985, p. 60.
Blewitt, Mike. SALES & MARKETING MANAGEMENT, July 1985, p. 76.
Bowers, Julie. MEETING NEWS, March 1989, p. 19.
Brass, Goldie. CANADIAN BUSINESS, Nov. 1986, p. 77.
Brennan, Leslie. SALES & MARKETING MANAGEMENT, Nov. 11, 1985, p. 96.
Brown, David. FORBES, Jan. 1, 1987, p. 108.
Brown, J. Michael. SALES & MARKETING MANAGEMENT, Nov. 11, 1985, p. 95.
Buckland, Leslie. BUSINESS WEEKLY, Aug. 11, 1986, p. 74.
CeSar, Don. MEETING NEWS, March, 1989, p. 21.
Chapman, Bryson. CANADIAN BUSINESS, Nov., 1986, p. 122.
Conway, Keith. Personal correspondence, July 15, 1989.
Corvo, David. WORKING WOMAN, June 1988, p. 42.
Crighton, Andrew. CANADIAN BUSINESS, Nov. 1986, p. 77.
Crump, John. BLACK ENTERPRISE, March, 1989, p. 40.
Daley, Kevin. WORKING WOMAN, Aug. 1987, p. 72.
Day, Richard. SALES & MARKETING MANAGEMENT, Nov. 11, 1985, p. 95.

Dovidio, John. FORTUNE, May 26, 1986, p. 178.
Draftz, John Z. Personal correspondence, July 10, 1989.
Drake, Mack. Personal correspondence, July 15, 1989.
Drucker, Peter. FORTUNE, July 11, 1983, p. 132.
Eaglin, Jim. BLACK ENTERPRISE, March 1987, p. 40.
Fisher, B. Aubrey. SMALL GROUP DECISION MAKING: COMMUNICATION AND GROUP PROCESS. New York: McGraw-Hill, 1974.
Fletcher, Winston. SAVVY, Nov. 1984, p. 29.
Frances, Richard. MEETINGS & CONVENTIONS, May 1988, p. 61.
Franke, Ralphe J. Personal correspondence, July 1, 1989.
Gackenbach, David E. ASSOCIATION MANAGEMENT, April 1985, p. 96.
Gillson, Ellen. Personal correspondence, June 4, 1989.
Green, Walter. U.S. NEWS AND WORLD REPORT, Oct. 12, 1987, p. 80.
Greenhalgh, Leonard. FORTUNE, May 26, 1986, p. 178.
Griffin, Dennis. Personal correspondence, June 30, 1989.
Grove, Andrew S. FORTUNE, July 11, 1983, p. 133.
Hall, Suzanne. SUCCESSFUL MEETINGS, Aug. 1988, p. 73.
Hamann, Joan. PERSONNEL ADMINISTRATOR, June 1986, p. 62.
Hamlin, Sonya. GLAMOUR, March 1988, p. 156.
Hanchette, Glynna. MEETING NEWS, March 1989, p. 19.
Hausman, Mike. SALES & MARKETING MANAGEMENT, July 1, 1985, p. 74.
Hemer, Ward. Personal correspondence, July 15, 1989.
Henze, Geraldine. WORKING WOMAN, Aug. 1987, p. 68.
Holbert, Edward B. ASSOCIATION MANAGEMENT, April 1985, p. 185.
Horiszny, Jean. PUBLIC RELATIONS JOURNAL, July 1986, p. 32.
Huffman, Carl. DATA MANAGEMENT, April 1987, p. 24.
Hyman, Beverly. SUCCESSFUL MEETINGS, March 1989, p. 107.
Janis, Irving L. GROUPTHINK: PSYCHOLOGICAL STUDIES OF POLICY DECISIONS AND FIASCOES. Boston: Houghton Mifflin, 1982.
Jeanguenin, Joanne C. BUSINESS MARKETING, Nov. 1985, p. 132.
Kelly, Joe. PERSONNEL, June 1987, p. 24.

Kiechell, Walter, III. FORTUNE, May 26, 1986, p. 177.
Kimes, James D. MANAGEMENT ACCOUNTING, Jan. 1987, p. 56.
Korda, Michael. POWER! HOW TO GET IT, HOW TO USE IT. New York: Random House, 1975.
Larson, Dale F. Personal correspondence, Aug. 1, 1989.
LeFaivre, Alvin. PUBLIC RELATIONS JOURNAL, Jan. 1987, p. 29.
Leigh, Andrew. INTERNATIONAL MANAGEMENT, May 1984, p. 22.
Lincoln, John A. SUPERVISION, Aug. 1988, p. 7.
Lindbloom, James E. Personal correspondence, Aug. 7, 1989.
Livingston-Mugnos, Linda. SALES & MARKETING MANAGEMENT, Nov. 1986, p. 117.
Lockshin, Jane. WORKING WOMAN, Aug. 1987, p. 69.
Lodge, Susan Guggenheim. WORKING WOMAN, Aug. 1987, p. 68.
Long, Janis K. Personal correspondence, May 25, 1989.
Lopez, Sylvia. Personal correspondence, June 2, 1989.
Lucas, Robert E. Personal correspondence, July 15, 1989.
Lutz, Charles. MEETINGS AND CONVENTIONS, Sept. 1988, p. 123.
McMahon, Tom. CANADIAN BUSINESS, Nov. 1986, p. 75.
Mackensen, Dan. Personal correspondence, June 10, 1989.
Milldebrandt, Tom. Personal correspondence, July 6, 1989.
Moore, Tom. ASSOCIATION MANAGEMENT, April 1985, p. 96.
Morrow, Laverne. BLACK ENTERPRISE, March 1987, p. 41.
Moskovitz, Helen. MEETINGS AND CONVENTIONS, May 1988, p. 289.
Mosvick, Roger K. WE'VE GOT TO START MEETING LIKE THIS! Glenview, IL: Scott, Foresman and Company, 1986.
Moynihan, Terry. CANADIAN BUSINESS, Nov. 1986, p. 78.
Munson, Robert J. Personal correspondence, June 18, 1989.
Munzer, Patti. WORKING WOMAN, Aug. 1987, p. 68.
Nestor, Paul. ASSOCIATION MANAGEMENT, April 1985, p. 99.
Nolan, Nancy N. PERSONNEL JOURNAL, Sept. 1986, p. 78.
Olenski, Michael. MARKETING NEWS, Sept. 25, 1987, p. 17.
Paris, Eileen. MEETING NEWS, March 1988, p. 19.
Parks, Ron. USAIR, Jan. 1987, p. 34.
Patterson, Jerry. Personal correspondence, July 7, 1989.
Pavlik, Nancy. MEETINGS & CONVENTIONS, Sept. 1988, p. 40.

Petersen, Marna. ASSOCIATION MANAGEMENT, April 1985, p. 94.

Picker, Scott. SALES & MARKETING MANAGEMENT, July 1, 1985, p. 78.

Pine, Donald L. Personal correspondence, June 8, 1989.

Pottash, Arthur. Personal correspondence, July 10, 1989.

Qubein, Nido. WORKING WOMAN, Aug. 1987, p. 68.

Redding, Charles. THE CORPORATE MANAGER'S GUIDE TO BETTER COMMUNICATION. Glenview, IL: Scott, Foresman and Company, 1985.

Reece, Beverly A. ASSOCIATION MANAGEMENT, April 1985, p. 97.

Reilly, Kate. BUSINESS MARKETING, Aug. 1987, p. 62.

Reynolds, Eleanor. WORKING WOMAN, Aug. 1987, p. 69.

Robertson, Dennis. MEETING NEWS, March 1988, p. 19.

Rondeau, Eugene. SALES & MARKETING MANAGEMENT, Nov. 1986, p. 94.

Rosenthal, Robert. SKILL IN NONVERBAL COMMUNICATION. Cambridge, MA: Oelgeschlager, Gunn, and Hain, 1979.

Rossman, Marlene. WORKING WOMAN, Aug. 1987, p. 68.

Schrage, Michael. *Wall Street Journal*, July 9, 1987, p. 2.

Scheef, Devon. Personal correspondence, July 15, 1989.

Schmidt, Gordon. Personal correspondence, July 25, 1989.

Sherick, Betty E. Personal correspondence, July 11, 1989.

Shrbeny, Ladislav. SUCCESSFUL MEETINGS, March 1989, p. 108.

Smith, Dayle. Personal correspondence, July 16, 1989.

Smith, Homer. SALES & MARKETING MANAGEMENT, Sept. 1986, p. 83.

Smith, Wayne. SUCCESSFUL MEETINGS, March 1989, p. 110.

Steinke, Richard. Personal correspondence, July 24, 1989.

Stoffmann, Daniel. CANADIAN BUSINESS, Nov. 1986, p. 75.

Stromp, Steven A. USAIR, Jan. 1987, p. 35.

Thompson, Ray. Personal correspondence, July 15, 1989.

Todd, Susan. Personal correspondence, June 11, 1989.

Toffler, Alvin. MEETINGS & CONVENTIONS, Sept. 1988, p. 24.

Tubbs, Stewart L. A SYSTEMS APPROACH TO SMALL GROUP INTERACTION. Reading, MA: Addison-Wesley, 1984.

Velasquez, Paul. Personal correspondence, July 30, 1989.

Wein, Cynthia. SUCCESSFUL MEETINGS, Dec. 1988, p. 72.

Wiegand, Richard. SUPERVISORY MANAGEMENT, July 1986, p. 38.
Wilkinson, Roderick. SUPERVISION, Oct. 1988, p. 15.
Will, Joe. CANADIAN BUSINESS, Nov. 1985, p. 122.
Wood, Julia T., et al. GROUP DISCUSSION: A PRACTICAL GUIDE TO PARTICIPATION AND LEADERSHIP. 2nd ed. New York: Harper and Row, 1986.
Wright, Philip. MEETING NEWS, March 1988, p. 18.
Wurtman, Judith. SUCCESSFUL MEETINGS, Aug. 1988, p. 74.

Index